Methods in Microanalysis

Volume II

WET COMBUSTION AND CATALYTIC METHODS IN MICROANALYSIS

Methods in Microanalysis

Volume II

WET COMBUSTION AND CATALYTIC METHODS IN MICROANALYSIS

Translated Microchemical Research Papers
of Professor Alexander Petrovich Terent'ev
Moscow University
Academy of Sciences, USSR

and of Jiři Körbl
Analytical Department, Research Institute
for Pharmacy and Biochemistry, Prague,
Academy of Sciences, Czechoslovakia

Edited by J. A. KUCK, Ph.D., *Senior Research Microanalyst*
American Cyanamid Company

Translated by KURT GINGOLD, Ph.D.(ATA), *Research Literature Scientist*
American Cyanamid Company

GORDON AND BREACH
Science Publishers
New York • London • Paris

150 Fifth Avenue
New York, N.Y. 10011

Library of Congress Card Number 64-18800

Editorial Offices for Great Britain and Europe:

Gordon and Breach Science Publishers, Ltd.
171 Strand
London, W.C.2, England

Distributed in the United Kingdom by:

Blackie & Son, Ltd.
5 Fitzhardinge Street
London W.1, England

PRINTED IN THE UNITED STATES OF AMERICA

A. P. TERENT'EV: On His 70th Birthday*

In 1961 there occurred the 70th birthday of one of the most prominent Soviet organic chemists, Aleksandr Petrovich Terent'ev, Corresponding Member of the Academy of Sciences of the U. S. S. R. since 1953, professor at Moscow University since 1935, member of the Commission on Analytical Chemistry. A student and co-worker of N. D. Zelinskiĭ, he has taken over in full measure the wide range of scientific interests and singular teaching skill of his teacher. Terent'ev has developed and introduced into practice a number of new synthetic methods, used for various classes of organic compounds.

A large place in Terent'ev's work is occupied by syntheses and studies of the properties of organic complexes, particularly inner complexes. He has made a particularly large contribution in the field of organic analysis.

Terent'ev's scientific creativity is characterized by a striving to expand the field of application of the new reactions and compounds studied by him. An important place is occupied here by the analytical use. Thus, after much work on the synthesis of diazo compounds, he proposed a method for the quantitative determination of compounds with conjugated double bonds, by coupling with diazotized *p*-nitroaniline

Zhur. Anal. Khim.* **16, 747-748 (1961)

A. P. TERENT'EV

(1936). This method is highly sensitive and permits the determination of conjugated dienes in the presence of a large excess of olefins.

After studying the structure and properties of organomagnesium compounds, Terent'ev made a fundamental improvement in the well-known Chugaev-Zerewitinoff method for determining active hydrogen. The Zerewitinoff method involves the difficult measurement of the volume of methane evolved from the expansion of the gas, and, in connection with this, specifies conducting the reaction not in ether, but in a high-boiling solvent.

Terent'ev, in 1940-1947, used for the Chugaev determination of active hydrogen a method of displacing the resulting methane with ether vapors, which substantially simplified the apparatus and method of determination, and permitted use of a more active ether solution of the organomagnesium compound, which in many cases reduced the time of the determination. The determination of active hydrogen by the Chugaev-Terent'ev method is widely used both in scientific investigations and in industrial control.

Terent'ev proposed an original method for the analysis of oleum, used widely in many laboratories.

In the first post-war years, Terent'ev and a group of co-workers finished developing a method for sulfonating organic compounds with complexed sulfur trioxide. This powerful synthetic method was also used by him for analytical purposes, for the quantitative determination of some forms of active hydrogen. For example, the action of a standard solution of sulfur trioxide on a hydroxy compound forms an alkylsulfuric acid, which is then diluted with water and titrated with alkali. The change in titre gives the amount of sulfur trioxide reacted, i. e., the equivalent of hydroxy compound. The method is unusual in that appreciable amounts of water do not interfere with the determination. Many other forms of active hydrogen and some functional groups are also determined in an analogous manner.

At the end of the 1930's, Terent'ev was much occupied with a study of the properties and reactions of acrylonitrile. Based on this, he and his co-workers developed a number of methods for the quantitative determination of acrylonitrile. The most interesting and effective method consists in the quantitative reaction of acrylonitrile with sodium sulfite in aqueous solution. Sodium hydroxide is formed, which can be titrated by any of the known methods. Having such a convenient method for the determination of acrylonitrile, Terent'ev, with his characteristic ingenuity, used it for the analysis of a number of cyanoethyl compounds. A sample of the compound is mixed with a standard acrylonitrile solution and, after some time, the unreacted acrylonitrile is back-titrated. This method has proved very convenient for the determination of thioalcohols, thiophenols, etc.

Terent'ev's strong imagination and unusual resourcefulness have enriched many branches of organic analysis. In the field of organic elemental analysis, he and his student, Dr. P. N. Fedoseev, have proposed methods for the decomposition of organic compounds by magnesium, calcium, and magnesium nitride.

In another group of papers, Terent'ev and B. M. Luskina developed a new, practical and valuable variant of the "wet combustion", permitting the simultaneous determination of carbon, halogen, nitrogen, and also ions of metals and silicon in the mineralized residue.

In the field of inorganic analysis, Terent'ev and co-workers have proposed a number of new organic reagents. Terent'ev's idea of "nascent reagents", developed recently by a number of authors, is interesting and fruitful. The precipitable ions and the precipitant ions are in solution not in the final form, but in a combined form: the inorganic ions in the form of unstable complexes, and the precipitant ions in the form of organic compounds. The latter hydrolyze to furnish the necessary anions to the solution. The slow formation of precipitant and precipitable ion favors good formation of crystals. For example, Terent'ev and co-workers proposed dimethyl sulfate, in the presence of sugar, for the precipitation of barium. Under these conditions, unusually large and easily filterable crystals of barium sulfate are formed. Under the same conditions, but in the presence of dioxane, calcium is precipitated as $CaSO_4 \cdot 0.5\ H_2O$, and can thus be determined in the presence of considerable quantities of strontium.

It would require a large monograph to characterize Terent'ev's numerous and various works in the field of analytical chemistry, but the significance of his activity in Soviet organic analysis is by no means limited to a few methods.

Terent'ev is rightly considered an outstanding Soviet organic analyst. This is not only because he has created a school of analysts and hundreds of his students are working successfully in various corners of our country. He is a tireless advocate of new, progressive methods of analysis, a severe critic, and an invaluable and benevolent consultant. A number of valuable analytical handbooks have been published in this country on his recommendation and under his editorship, and he has for many years been an active member of the Commission on Analytical Chemistry and of the editorial board of *Zhurnal Analiticheskoĭ Khimii*.

We wish him new great creative successes in all his varied scientific, public, and educational activities.

The Commission on
Analytical Chemistry

The Editors of *Zhurnal Analiticheskoĭ Khimii*

PREFACE

In keeping with our previously expressed interest in the current progress of Russian microchemical science, Volume II of this series presents useful material that has come to our notice as a result of our continuing study of Russian literature sources on the subject.

This time the central figure associated with many of the techniques under discussion in the book is still living. His long career is summarized in an introductory chapter which is included merely for the information of the reader. It indicates, nevertheless, how highly he is regarded as a chemist by his fellow countrymen.

Professor Terent'ev during his fruitful scientific career had a wide variety of interests in analytical chemistry, although he does not seem to have been especially concerned with microanalysis until within the last decade. We have therefore limited our selection of his research papers to those articles which are directly related to organic elemental analysis. These include: his techniques for decomposing organic compounds by means of magnesium or calcium or by magnesium nitride as substitutes for the traditional sodium fusion; his interesting series of twelve papers on the use of the "wet combustion" for analyzing refractory organic and organometallic structures; his use of the supplementary combustion over chromium sesquioxide at 750° to attack possible products of incomplete combustion that might escape the hot chromic acid solution; and his use of lithium aluminum hydride as a reagent in the active hydrogen determination.

Our survey of Professor Terent'ev's published research brought us inevitably into contact with the work of some of his students, since the latter upon their departure from Moscow University have often taken their research problems with them and have continued to work out their own ideas at their new locations. It would also appear that Terent'ev's influence has extended beyond the borders of his own country in view of the fact that his work is often closely associated with similar studies initiated at about the same time in several of the countries adjacent to the Soviet Union. Consequently we have included in Volume II articles by Czechoslovak authors as well as by several of Terent'ev's students whenever their papers were directly related to the subject under discussion.

As a result of this policy the content of Volume II can be broken down as follows: four chapters on fusion with alkaline earth metals; twelve chapters on the art of wet combustion; nine chapters on the determination of active hydrogen and microhydrogenation; three chapters on the use of chromium sesquioxide as an active combustion catalyst; and eleven chapters on the preparation, properties, and application of the thermal decomposition product of silver permanganate which is also an excellent combustion catalyst. Unfortunately there was not sufficient space in Volume II for the inclusion of material on the third important catalyst; i.e, cobaltous-cobaltic oxide, which belongs in the same category with the other two. Discussion of this important reagent, however, must be deferred to a later time.

May we also remind the reader that the present trend toward automation and speed-up in the microcombustion process calls for the most effective surface-combustion catalysts available. Therefore it is important for us as microchemists to implement our knowledge in this area.

To summarize: – Volume II of the series "Methods in Microanalysis" has a two-fold objective: 1) it endeavors to familiarize the reader with the possibilities of wet combustion in the microanalysis of difficult organic compounds and polymers with various combinations of heteroatoms; and 2) it presents valuable information and useful laboratory directions in connection with catalytic oxidation and reduction in microanalysis. It is hoped that those analysts, finding it difficult to get complete disintegration with certain of their refractory samples by their present methods, may perhaps find useful suggestions in the present work.

J. A. Kuck

TRANSLATOR'S PREFACE

Aside from the usual difficulties that a translator must overcome if he is to succeed in his attempt to provide an accurate, yet readable, English version of a foreign work, there are two special problem areas that might be mentioned in connection with the present series of translated papers.

First, the matter of organic nomenclature: The systems used by Russian authors often differ considerably from those in use in this country. Furthermore, in some cases, the names given are not specific and, unless accompanied by structural formulas, cannot be identified with certainty. In the translations, every effort has been made to use names that will be intelligible to English-speaking readers wherever possible. In this connection, it is my pleasant duty to acknowledge the valuable contribution made by Miss Christel J. Kappes, formerly Research Literature Scientist with American Cyanamid Company, who helped us cheerfully and willingly with questions of nomenclature.

The second problem relates to the bibliographies at the ends of chapters. In many cases, the names of authors or journals, or the page references, are given incorrectly in the papers translated here. I have attempted to check all references against the original articles cited, where the journals in question were available to me, or against *Chemical Abstracts* or

Chemisches Zentralblatt where they were not, and I hope that this service -- which is not generally part of the translator's function -- will prove of value to users of this series.

Kurt Gingold

CONTENTS

1

The Use of Alkaline Earth Metals in Organic Elemental Analysis *

A. P. Terent'ev, P. N. Fedoseev,
and
N. P. Ivashova

The determination of nitrogen, sulfur, and halogens is very important in the practice of organic chemical analysis. Among the many methods that have been used in the determination of these elements are methods based on the decomposition of organic compounds by active metals. The use of alkali and alkaline earth metals in analysis attracted investigators a long time ago [1, 2]. Alkali metals are used particularly frequently for these purposes [2-5]. However, working with these involves a number of undesirable features related to their properties.

In our work, we have used the powdered alkaline-earth metals magnesium and calcium for the decomposition of organic compounds, using their ability to extract nitrogen, sulfur, and halogens readily from compounds containing these elements. The use of alkaline-earth metals in analysis makes it possible not only to carry out separate determinations of nitrogen, sulfur, and halogens, but also to determine two or three of these elements simultaneously in a single

* *Izvest. Vysshikh Ucheb. Zavedeniĭ, Khim. i Khim. Tekhnol.* **2**, *No.*1, 54-58 (1959)

sample, which is one of the significant advantages of the magnesium and calcium methods. Our methods have been used to develop a quantitative micro method for the determination of nitrogen, sulfur, and halogens in organic compounds [6].

The principle of these methods for the qualitative and quantitative determination of nitrogen, sulfur, and halogens in organic compounds is the decomposition of the analytical sample by heating with powdered magnesium (calcium). In the determination of nitrogen, the compounds are decomposed in the absence of air, since atmospheric nitrogen, on heating, reacts with magnesium (calcium), leading to a high value for nitrogen. Magnesium fixes the nitrogen of organic compounds as magnesium nitride, sulfur as sulfide, and halogen as magnesium halide. Substitution of calcium for magnesium gives calcium nitride, sulfide, and halide. The use of calcium permits a qualitative determination of carbon in organic compounds, in the form of calcium carbide.

Despite the fact that the principle for the determination of elements is general for the magnesium and calcium methods, there are a number of substantial differences in the methods of determination. This permits using both metals, depending on the working conditions [7].

The melting point of calcium is 850°; that of magnesium 650°. This makes it possible to decompose a sample with calcium at 750-800°, and with magnesium at 600-640°. Unlike magnesium, calcium does not react with glass at the temperature in question, which considerably simplifies the method of analysis.

Owing to its high activity, calcium fixes all of the gaseous decomposition products of the sample formed in the combustion tube, and produces a vacuum in the tube. (At 700°, the tube collapses and closes by fusion.) This ensures against sample loss. Replacement of magnesium by calcium permits a slight simplification of the technique and a decrease in analysis time. But calcium is less available than magnesium; also, it changes in air and must therefore be stored under exclusion of air (in ground-glass-stoppered jars in an inert gas atmosphere) and must be handled with due precautions.

The advantages of magnesium are its stability to external influences, low specific gravity (less magnesium than calcium is required for each analysis), safe handling, and availability in any laboratory.

The purpose of the present paper is to describe methods for the qualitative determination of nitrogen, sulfur, and halogens in organic compounds by heating with metallic magnesium or calcium, and for the determination of carbon by using metallic calcium for decomposition of the sample. Some characteristics of the calcium and magnesium methods for the quantitative determination of elements in organic compounds are noted in the paper.

EXPERIMENTAL PART

QUALITATIVE DETERMINATION OF NITROGEN, SULFUR, AND HALOGENS IN ORGANIC COMPOUNDS BY HEATING WITH POWDERED MAGNESIUM.

1-2 granules (1-5 mg) of the organic sample are placed in a small glass tube (length 80-100 mm, diameter 3-5 mm), sealed at one end, which has a small amount of magnesium (5-10 mg) at its bottom.

The sample is mixed with the magnesium, and another 50-60 mg of magnesium is introduced into the tube, and is distributed as a uniform layer reaching halfway up the tube. Air is displaced from the tube with 3-5 drops of ether. The open end of the tube is closed with a rubber tube, whose end is immersed in water. The air in the rubber tube is also displaced with ether (the air need not be displaced in the determination of sulfur and halogens). To prevent water from being sucked in, a Bunsen valve is placed on the open end of the rubber tube immersed in the water. The magnesium in the tube is heated in the flame of a gas burner in the direction towards the sample, so that the vapors of the sample pass through red-hot magnesium.

After the whole tube has been heated to red heat (600-650°), it is cooled, wrapped in glossy paper, broken in the paper, and the contents are poured into a dry test tube. The reaction mass in the test tube is treated with 2-3 ml of water, the solid residue is filtered off, and the desired elements are determined in the filtrate.

To determine sulfur, 0.1-0.2 ml of 2 N ammonia and 0.1-0.2 ml of a 1% sodium nitroprusside solution are added to one third of the filtrate. The appearance of a pink or pink-violet coloration indicates the presence of sulfur.

The remaining two thirds of the filtrate are acidified with a few drops of dilute sulfuric acid and boiled to remove hydrogen sulfide. The solution obtained is divided into two parts. In one part of the solution, nitrogen is determined with Nessler reagent. The presence of nitrogen is shown by the appearance of a brown coloration or a brown-red precipitate when 0.1 ml of Nessler reagent is added to part of the solution. In the second part of the solution, halogens are determined by the addition of 0.1 ml of 0.1 N silver nitrate solution.

QUALITATIVE DETERMINATION OF CARBON, NITROGEN, SULFUR, AND HALOGENS IN ORGANIC COMPOUNDS BY HEATING WITH POWDERED CALCIUM.

1-2 granules (1-5 mg) of the organic sample are placed in a glass tube (length 80-100 mm, diameter 3-5 mm) sealed at one end, which has 10-15 mg of calcium at its bottom. The sample is mixed with the calcium, and another 80-100 mg of calcium is introduced, and is distributed as a uniform layer reaching half-way up the tube. Air is removed from the glass tube by means of powdered calcium. This is done by placing 70-90 mg of powdered calcium into a quartz or porcelain tube (length 60-80 mm, diameter 3-5 mm) and connecting this to the analysis tube by means of a 30-40 mm length of rubber vacuum tubing. The calcium in the quartz tube is initially heated to 700-800° for 5-7 minutes. The heated calcium fixes the air contained in the tubes. The tube with the sample is then heated as described for magnesium. The air need not be removed when determining sulfur, halogens, and carbon.

After the whole sample tube has been heated to red heat, it is cooled and disconnected from the rubber tubing. The analysis tube is wrapped in glossy paper, broken up in the paper, and the contents are poured into a dry test tube 1. The test tube 1 is connected to test tubes 2 and 3. Test tube 2 contains 1 ml of saturated lead acetate solution, and test tube 3, 1 ml of Ilosvay reagent. *

The reaction mass in test tube 1 is treated with 1-2 ml of 5% acetic acid, added from a dropping funnel. The liberated acetylene is absorbed in test tube 3. The presence of carbon in the sample is shown by the formation of a dark red precipitate of copper acetylide. If the sample contains a significant amount of sulfur, a black precipitate of lead sulfide is formed in test tube 2.

After complete decomposition of the reaction mixture in test tube 1, the solution is poured from 1 into a clean test tube. Nitrogen and halogens are then determined in it, and a test is made for the presence of sulfur. This is done by dividing the liquid into three parts and proceeding as described in the case of magnesium.

We have analyzed various classes of organic compounds: toluidine, *o*-nitrophenol, chloroacetic acid, aminosulfopyridine, diphenylthiourea, 5,7-dibromohydroxyquinoline, sulfosalicylic acid, hexachloroethane, benzothiazole sulfide, dimethylamine, etc. The nitrogen content in these ranged from 4-19%, sulfur 12-15%, halogens 30-90%, and carbon 10-80%. In all of these cases, distinct reactions were obtained for nitrogen, sulfur, halogens, and carbon. As a result of these analyses, we have established the minimum amounts detectable in the determination of the individual elements with magnesium and calcium. (Table 1).

---

* Ilosvay reagent is prepared by dissolving 1 g of copper nitrate in 4 ml of ammonia, adding 3 g of hydroxylamine hydrochloride, and making up to 50 ml with water.

TABLE 1

Minimum detectable amounts in the determination of individual elements in organic compounds with magnesium and calcium.

Element	Minimum detectable amount (μg)	
	Mg method	Ca method
Nitrogen	18	30
Sulfur	35	60
Halogen	110	155
Carbon	-	300

QUANTITATIVE DETERMINATION OF NITROGEN, SULFUR, AND HALOGENS IN ORGANIC COMPOUNDS.

The decomposition of the sample is carried out in a small glass tube sealed at one end. 50-70 mg of sample is placed at the bottom of the tube and mixed with a small amount (50-100 mg) of magnesium or calcium. To avoid the possible escape of sample from the tube, excess metal (e.g., 1 g of magnesium or 1.5-2 g of calcium) is placed on top of the sample, and air is displaced from the tube. The air can be displaced with hydrogen or ether, but, when calcium is used in the analysis, it is convenient to absorb the air with metallic calcium contained in a separate tube.* The sample tube is closed with a gas outlet tube, whose end is immersed in water. The

* If the object of the analysis is to determine sulfur or halogens, the air is not displaced.

heating of the tube is started at the end opposite the sample, and is continued in a direction towards the sample, gradually expanding the zone of heated metal. After heating, the cooled tube is wrapped in a clean filter paper, broken, and transferred together with the paper into a flask for distillation of ammonia and hydrogen sulfide. When using magnesium, the reaction mass in the flask is treated with sulfuric acid, after first displacing the air from the apparatus with carbon dioxide.

To determine nitrogen, the acid solution is made alkaline, and the liberated ammonia distilled into a standard solution of mineral acid or potassium biiodate (KHI_2O_6). If calcium was used in the analysis, the reaction mass in the flask is decomposed with distilled water, and the air is not displaced from the apparatus. Ammonia is distilled from the alkaline solution, and the nitrogen content is determined acidimetrically or iodometrically.

The results of some analyses are given in Table 2.

The methods for determining sulfur and halogens are the same in the calcium and magnesium analyses. To determine sulfur, air is displaced from the flask with carbon dioxide, the reaction mass is decomposed with acid, and the hydrogen sulfide liberated is absorbed in a receiver with a solution of zinc or cadmium acetate. The sulfur content in the sample is determined iodometrically. To determine halogens, the reaction mass is treated with sulfuric acid until the metal is completely dissolved, solid residues are filtered off, and the halogen content is determined argentometrically.

In the simultaneous determination of two or three elements, sulfur is first determined by distillation of hydrogen sulfide from the acid solution; the

TABLE 2

Results of the quantitative determination of nitrogen, sulfur, and halogen in some organic compounds with magnesium and calcium.

Sample No.	Name of Compound	mg Sample	%N			%S			%X		
			Calc.	Found	△	Calc.	Found	△	Calc.	Found	△
1	α-Phenyl-N-methylphenylene-thiazine perchlorate	72.1	4.12	3.98	-0.14	9.44	9.21	-0.23	10.43	10.35	-0.08
		73.1		3.91	-0.21		9.26	-0.18		10.28	-0.15
2	Sulfanilic acid	34.2	6.69	6.47	-0.22	15.32	15.12	-0.20	--	--	--
		39.7		6.60	-0.09		15.14	-0.18			
3	Acridine	48.2	7.81	7.68	-0.13	--	--	--	--	--	--
		51.2		7.86	+0.05						
4	Aminopyridine	60.9	29.76	29.63	-0.13	--	--	--	--	--	--
		39.8		29.56	-0.20						
5	Hexachloroethane	34.5	--	--	--	--	--	--	89.85	89.92	+0.07
		39.1								89.65	-0.20
6	Mercaptobenzo-thiazole	37.2	8.35	8.26	-0.09	38.31	38.40	+0.09	--	--	--
		46.0		8.15	-0.20		38.08	-0.23			

Note: Samples 1, 2, and 4 were analyzed with magnesium; samples 3, 5, and 6 with calcium.

solution is then made alkaline and the nitrogen is determined as ammonia. Halogen is determined in the residue in the distillation flask by argentometric titration.

SUMMARY

1. Methods for the qualitative determination of nitrogen, sulfur, and halogens in organic compounds by means of magnesium and calcium are described.

2. The use of calcium permits the qualitative determination of four elements: carbon, nitrogen, sulfur, and halogen.

3. The advantages of the calcium method over the magnesium method in quantitative analysis are shown.

4. However, despite several advantages of the calcium method, the wide availability of metallic magnesium, the convenience of working with it, and its low specific gravity must promote the use of the magnesium method along with the calcium method.

LITERATURE CITED

[1] V. Castellana, *Gazz. Chim. Ital.* **34II**, 357 (1904); J. F. Thorpe and M. A. Whiteley: "Laboratory Manual of Organic Analysis" [Russian Edition], Moscow (1937).

[2] F. Emich: "Microchemical Analysis" [Russian Edition], Leningrad (1932).

[3] F. Beilstein, *Ber.* 5, 620 (1872).

[4] A. P. Lidov: "Gas Analysis", Leningrad (1918);

A. P. Stepanov, *Zhur. Russ. Fiz.-Khim. Obshchestva* **37**, 12 (1905); **39**, 4056 (1906).

[5] M. O. Korshun and N. E. Gel'man: "New Methods of Elemental Microanalysis", Moscow-Leningrad (1949); K. Bürger, *Angew. Chem.* **54**, 479 (1941); *Die Chemie* **55**, 245 (1942); W. Zimmermann, *Mikrochemie ver. Mikrochim. Acta* **31**, 15 (1943); **33**, 122 (1947).

[6] W. Schöniger, *Mikrochim. Acta* **1954**, 74; **1955**, 44.

[7] P. N. Fedoseev and N. P. Ivashova, *Zhur. Anal. Khim.* **7**, 112, 116 (1952); **11**, 233 (1956); A. P. Terent'ev. P. N. Fedoseev, and N. P. Ivashova, *ibid.* **13**, 344 (1958).

2

The Qualitative Determination of S, X, C, P, As, Sb and Bi in Organic Compounds with Magnesium Nitride*

A. P. Terent'ev, S. I. Obtemperanskaya, and V. A. Dolgikh

We have described a method for the quantitative determination of chlorine and bromine in organic compounds, based on the decomposition of the sample with magnesium nitride. [1] It has been found that magnesium nitride (Mg_3N_2) can also be used for the qualitative determination of a number of elements in organic compounds, including sulfur, phosphorus, arsenic, antimony, bismuth, carbon, chlorine, bromine, and iodine.

The determination is based on the reductive decomposition of the organic compound with magnesium nitride at 650-800°. The halogen forms magnesium halide; sulfur is partly liberated as hydrogen sulfide and partly converted to magnesium sulfide; arsenic antimony, and bismuth are partly liberated in the elemental state on the walls of the test tube and partly form arsenide, antimonide, and bismuthide; phosphorus gives magnesium phosphide. On decomposition of the residue with water (after calcination) and aci-

* *Nauch. Doklady Vyssheǐ Shkoly, Khim i Khim. Tekhnol.* **1959**, *No.* 2, 305-306.

dification of the solution, carbon is liberated as charcoal or carbon black.

The proposed method makes possible the simultaneous detection of the above elements in a substance or mixture of substances.

EXPERIMENTAL PART

Magnesium nitride is prepared by heating powdered magnesium metal in a porcelain crucible, closed with a perforated iron cover, in a muffle furnace for 20-30 minutes at 700-800°. [1]

A small amount of the substance to be analyzed is placed in a hard glass test tube (length 22 cm, diameter 1 cm), covered with 0.5-1g of Mg_3N_2, and heated to red heat for 5-10 minutes in a gas burner. The heating is started at the open end, so that the vapors of the sample pass through a red-hot zone of Mg_3N_2.

To detect chlorine, bromine and iodine, the test tube is cooled, the contents are poured into an Erlenmeyer flask containing 3-5 ml of distilled water, acidified with 20% HNO_3, the solution is filtered, and the filtrate is tested for the presence of halide ion with $AgNO_3$ solution. Halogen was detected by this method in iodotoluene, iodoform, bromobenzene, dibromobenzene, bromobenzoic acid, bromopyromucic acid, bromonitrobenzene, dichloroquinone, chloroquinone, chlorobenzene, trichloroacetic acid, and chloral hydrate.

The minimum detectable amounts are: 0.1 mg for chlorine and bromine, 1 mg for iodine.

Sulfur is detected in most cases by the evolution of H_2S, which is identified by the blackening of a filter paper moistened with lead acetate, or by the appearance of a violet color on a paper moistened with sodium nitroprusside. If the result is doubtful, the test tube is cooled, the contents are poured into an Erlenmeyer flask containing 3-4 ml of water, the solution is filtered, acidified with acetic acid, and tested for sulfide ion by adding a 10% solution of lead acetate or sodium nitroprusside. Sulfur was detected by this method in β-thionaphthol, dibenzyl sulfide, sulfanilic acid, 4-methyl-2-aminothiazole, thioxane, sulfidine, streptocide, benzyl mercaptan, and β-naphthalenesulfonic acid. The minimum detectable amount is 5 mg.

To detect phosphorus, the test tube is cooled, the contents are poured in portions into an Erlenmeyer flask containing 5 ml of distilled water, and 5 ml of concentrated HNO_3 is added. The solution is filtered, excess ammonium molybdate solution is added, it is heated to 50-60°, and left to stand for a while. The formation of a yellow, crystalline precipitate of ammonium phosphomolybdate indicates the presence of phosphate ion. The minimum detectable amount is 20 mg. Phosphorus was detected by this method in diethyl phosphoric acid, triethylphosphine, and triethyl phosphite.

In the detection of arsenic, antimony, and bismuth, a black deposit of arsenic (or antimony or bismuth) is formed on the cold part of the test tube. The test tube is cooled, the contents are transferred to a 50-ml round-bottomed flask provided with an outlet tube, the flask is closed with a stopper carrying a dropping funnel, the outlet tube is immersed in a test tube containing 0.1 N $AgNO_3$ solution, and distilled water, acidified with HCl, is added to the flask from the dropping funnel.

Arsine (or the corresponding hydrides of antimony or bismuth) is liberated and is detected by the appearance of a black precipitate of silver in the test tube (in a hood). The minimum detectable amount is 0.1 mg for arsenic, 10 mg for antimony and bismuth. Arsenic was detected by this method in triphenylarsine oxide, chlorodiphenylarsine, dichlorophenylarsine oxide, and phenylarsenic acid, and antimony in dichlorotriphenylantimony, acetyldiphenylantimony, tritolylantimony, and diphenylantimonic acid.

To detect carbon, the test tube is cooled, the contents are poured into an Erlenmeyer flask containing 5 ml of distilled water, and acidified with 20% HNO_3 until an acid reaction is obtained. Carbon is liberated in the free state as carbon black. For comparison, a blank is run with the same amount of Mg_3N_2. The minimum detectable amount is 5 mg. Carbon was detected by this method in succinic acid, naphthalene, sodium acetate, sodium formate, and in all the organic compounds that had been analyzed for the presence of halogen, sulfur, and other elements.

The same methods can be used when a simultaneous detection of these elements in a substance or mixture of substances is required.

LITERATURE CITED

[1] A. P. Terent'ev, S. I. Obtemperanskaya, and N. V. Ermolenko, *Nauch. Doklady Vyssheĭ Shkoly, Khim. i Khim. Tekhnol.* 1958, 83.

3

The Determination of Chlorine and Bromine in Organic Compounds with Magnesium Nitride*

A. P. Terent'ev, S. I. Obtemperanskaya,
and
N. V. Ermolenko

Many methods are known for determining halogen in organic compounds. The recently developed reductive methods, based on fusion of the compound being analyzed with metallic potassium [1, 2] or with powdered magnesium [3, 4, 5], are sufficiently simple and accurate, but have not yet found wide application, since they have certain inconveniences and shortcomings. Thus, for example, in the Fedoseev-Ivashova method [4], the reaction mass, after ignition with metallic magnesium, cannot be extracted from the test tube, because of fusion with the glass, and the test tube must be broken and ground up in a mortar. Moreover, when the reaction product, which may contain silicides formed in the fusion, is acidified, simultaneous formation of silane and a detonating mixture is possible. Ignition of the silane in air may lead to explosion of the detonating mixture.

The simple, easy to perform, safe, and rapid method for determining chlorine and bromine in organic compounds now proposed by us is based on the reductive decomposition of the organic chlorine-or

* *Nauch. Doklady Vyssheĭ Shkoly, Khim. i. Khim. Technol.* 1958, *No.* 1, 83-85.

bromine-containing compound by magnesium nitride, followed by a determination of the halide ion in the resulting magnesium halide by the argentometric Volhard titration. We did not obtain quantitative, reproducible results for the determination of iodine.

The decomposition of the weighed sample of chlorine- or bromine-containing organic compound was carried out in long, open, hard-glass test tubes, by heating in the roaring flame of the gas burner. The reaction is complete in 5-10 minutes for micro samples.

The reaction mass did not fuse with the glass and was easily extracted from the test tube. No explosions were observed when the reaction product was acidified.

PREPARATION OF MAGNESIUM NITRIDE

A porcelain crucible was three-quarters filled with magnesium powder, closed with an iron cover containing small openings, and kept for 20-30 minutes in a muffle furnace at 700-800°. After cooling, the white upper layer of magnesium oxide is removed; the yellow lower layer of magnesium nitride is ground in a mortar and stored in a ground-stoppered container.

QUANTITATIVE DETERMINATION OF CHLORINE AND BROMINE IN ORGANIC COMPOUNDS SEMIMICRO METHOD

Reagents: Magnesium nitride; nitric acid (sp. gr. 1.4); 0.1 N silver nitrate; 0.1 N ammonium thiocyanate; phenolphthalein solution (1 g indicator, 70 ml ethanol, and 30 ml water); saturated solution of ferric ammonium alum.

A small amount (about 1 g) of magnesium nitride is placed in a hard-glass test tube (length 22 cm, diameter 1 cm), and 20-70 mg of sample is added. Magnesium nitride is added until the tube is about half-full (about 4 g).

Heating is carried out in the roaring flame of the gas burner, starting at the open end. so that the vapors of the sample pass through the incandescent zone of magnesium nitride. The burner is gradually moved towards the sample, with a speed depending on the nature of the compound under analysis. If the decomposition of the sample is too violent, the burner should be moved back a little towards the open end, but should not be removed altogether or moved beyond the layer of magnesium nitride. After carbonization of the sample and cessation of gas evolution, heating is continued for another 30 minutes at red heat.

The test tube is cooled and the contents are poured in portions into an Erlenmeyer flask containing 25 ml of distilled water. The flask is cooled with water during the addition. 65% Nitric acid is added until the solution is acid to phenolphthalein, the black precipitate of carbon is filtered off, and halide ion is determined in the filtrate by the Volhard method. The microdetermination of chlorine and bromine (3-7 mg of sample) is carried out by the same method, using shorter and narrower test tubes, a smaller layer of magnesium nitride (3 cm), 0.02 N silver nitrate and ammonium thiocyanate, and decreasing the time for ignition of the sample with magnesium nitride to 5-10 minutes.

The results of the determinations are shown in Tables 1 and 2.

The error in the determination of chlorine and bromine by the proposed method thus does not exceed ±0.3%.

TABLE 1

Semimicro Method

Compound	g Sample	%X Calc.	%X Found	Δ
2, 3-Dichlorobenzoquinone $C_6H_2Cl_2O_2$	0.0400	40.07	39.89	-0.18
	0.0496		40.03	-0.04
Chlorobenzene C_6H_5Cl	0.0748	31.50	31.29	-0.21
	0.0409		31.21	-0.29
Chloral hydrate $C_2H_3Cl_3O_2$	0.0237	64.31	64.20	-0.11
	0.0315		64.16	-0.15
Chlorobenzoquinone $C_6H_3ClO_2$	0.0545	24.88	24.72	-0.16
	0.0455		24.93	+0.05
α-Bromopyromucic acid $C_5H_3BrO_3$	0.0422	41.84	41.66	-0.18
	0.0461		41.60	-0.24
p-Bromobenzoic acid $C_7H_5BrO_2$	0.0504	39.76	39.64	-0.12
	0.0391		39.85	+0.09
p-Dibromobenzene $C_6H_4Br_2$	0.0634	67.77	68.01	+0.24
	0.0625		67.77	0.00
p-Bromonitrobenzene $C_6H_4BrNO_2$	0.0543	39.56	39.74	+0.18
	0.0565		39.60	+0.04
Bromobenzene C_6H_5Br	0.0709	50.90	50.72	-0.18
	0.0408		50.98	+0.08

TABLE 2

Micro Method

Compound	mg Sample	%X		Δ
		Calc.	Found	
Chlorobenzoquinone	3. 220	24. 88	24. 89	+0. 01
$C_6H_3ClO_2$	3. 340		24. 73	-0. 15
Chloral hydrate	3. 215	64. 31	64. 18	-0. 13
$C_2H_3Cl_3O_2$	3. 500		64. 12	-0. 19
p-Bromobenzoic acid	4. 810	39. 76	39. 88	+0. 12
$C_7H_5BrO_2$	3. 010		39. 80	+0. 04
p-Dibromobenzene	3. 540	67. 77	67. 73	-0. 04
$C_6H_4Br_2$	4. 120		67. 86	+0. 09

LITERATURE CITED

[1] M. O. Korshun and N. E. Gel'man: "New Methods of Elemental Microanalysis", Goskhimizdat (1949).

[2] M. O. Korshun and M. N. Chumachenko, *Dokl. Akad. Nauk S.S.S.R.* **99**, 769 (1954).

[3] P. N. Fedoseev and N. P. Ivashova, *Zhur. Anal. Khim.* **7**, 112 (1952).

[4] P. N. Fedoseev and N. P. Ivashova, *ibid.* **11**, 233 (1956).

[5] W. Schöniger, *Mikrochim. Acta* **1954**, 74.

4

The Simultaneous Determination of N, S and X by Means of Calcium *

A. P. Terent'ev, P. N. Fedoseev,
and
N. P. Ivashova

In earlier papers we have shown that nitrogen, sulfur, and halogen in organic compounds can be determined quantitatively by means of powdered metallic magnesium. [1] The disadvantages of this method include the need for treating the reaction mixture with acid, leading to the formation of explosive mixtures of silanes, hydrogen, and air, which is why the decomposition is carried out in an inert gas atmosphere. In order to simplify the magnesium method and to eliminate the disadvantages cited, we decided to use powdered calcium instead of magnesium for the determination of nitrogen, sulfur, and halogen in organic compounds.

Calcium already reacts with water at ambient temperature, and combines readily with oxygen, nitrogen, sulfur, and halogens. The ability of calcium, on heating, to fix traces of air and other gases and vapors is used in technology to produce vacuum. Unlike magnesium, calcium does not react with glass at 700-750°. The m. p. of calcium is 850°; its b. p. 1487°.

* *Zhur. Anal. Khim.* **13**, 344-348 (1958).

Magnesium melts at 650° and boils at 1100°.

The principle of the method consists in heating a sample of the organic compound under analysis containing nitrogen, sulfur, and halogen, with powdered calcium at 700-750° in an atmosphere of ether vapor or hydrogen. Calcium nitride, sulfide, and halide are formed, in which nitrogen, sulfur, and halogen are determined quantitatively.

DETERMINATION OF NITROGEN

The sample is decomposed in the apparatus shown in Fig. 1. 30-60 mg are introduced into the bottom of the glass tube 1 with a thin metal spatula.* The weighed sample is mixed with 20-30 mg of powdered calcium **, the mixture is stirred, and another 1-1.5 g of calcium is placed on top. The calcium layer in the tube should occupy 100-120 mm of the tube length. Air is displaced from the tube by pouring 1 ml of ether into it.

To the open end of the tube is attached a rubber tube 2, connected to tubes 4 and 3. A Bunsen valve is inserted in tube 3, which is passed under a cylinder of water. [Ed. note: Presumably the tubing is closed with a dead end and a slit cut in the rubber.] Air is also removed from tubes 2, 3, and 4 by pouring 1.5 ml of ether into them. Tube 1, prepared for analysis,

*The spatula is a copper wire (length 300 mm, diameter 3 mm); one end is flattened to a small spatula, convenient for introducing the sample into the tube.

**The calcium cannot be stored in air for any length of time; it must be kept in glass jars with ground stoppers, filled with CO_2.

is placed into a hinged tubular electric furnace 8 heated to 700-750°, or is heated with gas burners. Heating is started at the calcium layer furthest from the sample, and the heating zone is gradually expanded in the direction of the sample. The heated calcium fixes all the gaseous decomposition products formed in the tube. This can be seen from the sucking in of water and the collapse of the glass tube 1. To prevent water from getting into the analysis tube, the rubber tube 2 is closed with a clamp at the point of attachment to tube 1. The clamp is closed at the instant when the water sucked in appears in the connecting glass tube 4.

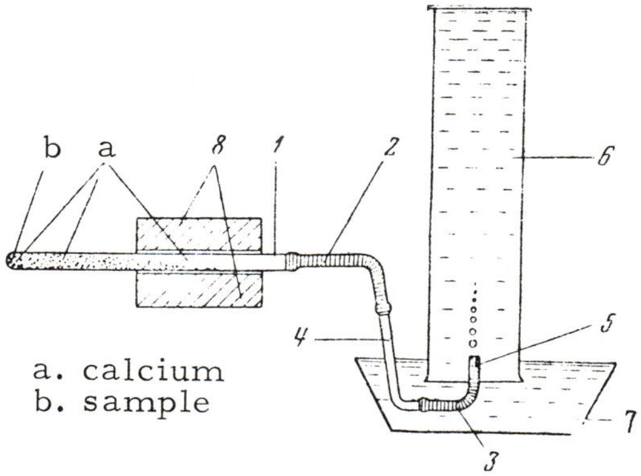

Fig. 1 Decomposition apparatus. 1: Glass tube (length 250 mm, i.d. 5-6 mm), sealed at one end; 2 and 3: rubber tubes (length 200 mm, diameter 4-5 mm); 4: glass tube; 5: Bunsen valve; 6: cylinder containing water; 7: crystallizing basin.

After all of the calcium layer has been heated to the stated temperature (15-20 minutes), the electric furnace is turned off. A vacuum must be maintained in the cooling tube (as seen by the collapsed rubber), otherwise the analytical results for nitrogen may be too high. The decomposition of an organic sample by heating with calcium may also be carried out in another variant of the apparatus, shown in Fig. 2, which permits dispensing with the displacement of air by ether or hydrogen.

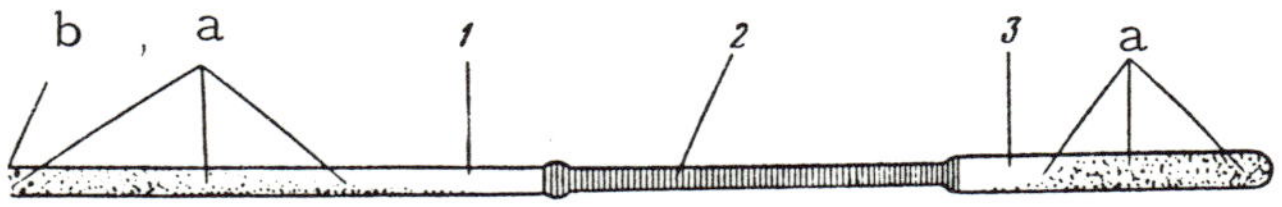

Fig. 2 Simplified decomposition apparatus. 1: Glass tube (described above); 2: vacuum rubber tubing (length 50-60 mm); 3: quartz or porcelain tube (length 150 mm, diameter 6-7 mm), sealed at one end; a: calcium; b: sample.

Tube 1 is prepared for analysis as indicated above, and the sample and powdered calcium are introduced into it. To remove air from tube 1, 200-300 mg of powdered calcium is placed in the quartz tube 3, and tube 1 is connected to tube 3 by means of vacuum rubber tubing 2.

The calcium in the quartz tube 3 is first heated to 800°. Within 20 minutes, the heated calcium absorbs all the air contained in the apparatus. After this, tube 1 with the sample is heated as described above. The cooled tube 1 is removed from the furnace, the part of the tube not containing calcium is cut off and discarded, and the remainder of the tube, containing the sample and calcium, is wrapped in a filter paper, crushed with a hammer, and transferred together with the paper to the dry round-bottomed flask of an ammonia-distillation apparatus. 20 ml of 0.05 N mineral acid and 2 drops of indicator (methyl red) are placed in the receiving flask. 80-100 ml of distilled water is added to the flask from a dropping funnel to decompose the reaction mass.* Since the calcium reacts vigorously with water, the water is added dropwise at first. When the vigorous reaction ceases, the rest of the water is added. The flask is heated and ammonia is distilled off (20-25 minutes). The excess acid in the receiver is back-titrated with 0.05 N alkali.

The results of some determinations in this series of experiments are given in the table.

DETERMINATION OF SULFUR

The sample is decomposed in the apparatus shown in Fig. 1, which we described for the determination of nitrogen. The sample is introduced into the glass tube 1, and 1-1.5 g of powdered calcium is added (see Determination of Nitrogen).**

Tube 1 is prepared for analysis, and the apparatus is assembled. (Fig. 1)

Tube 1 is placed in a hinged tubular electric furnace and heated as described for the determination of nitrogen. When heating is completed, the furnace is turned off, the tube is removed, cooled, wrapped in a

*The decomposition may also be carried out with 30% ethanol. In this case, the reaction proceeds less energetically.

**In the determination of sulfur, it is unnecessary to remove air from the apparatus.

piece of filter paper, crushed, and transferred together with the filter paper to the dry, round-bottomed distillation flask of an apparatus for distillation of H_2S. The apparatus contains three absorption flasks for absorbing H_2S. The first absorption flask contains 25-30 ml of a mixture of 10% sodium acetate solution and 10% zinc sulfate solution, acidified with a few drops of concentrated acetic acid. The second and third flasks contain, respectively, 20 and 10 ml of the same mixture. A small excess of 4% H_2SO_4 or HCl is added to the distillation flask slowly (particularly at first) from a separatory funnel, until the calcium is completely dissolved and the CaS completely decomposed. After the addition of acid is completed, the flask is heated and H_2S is distilled off together with liberated hydrogen. Distillation is stopped when one third of the flask's content has distilled. H_2S is absorbed in the form of ZnS, mainly in the first receiver.

When distillation of H_2S is completed, the contents of all three receivers are poured together, 20 ml of 0.05 N iodine is added, acidified with a few drops of 10% HCl, and, after standing for 3 minutes, excess iodine is back-titrated with 0.05 N $Na_2S_2O_3$. The results of some determinations are given in the table.

DETERMINATION OF HALOGEN

The determination is carried out in the apparatus shown in Fig. 1. The weighed sample and 1-1.5 g of powdered calcium are introduced into tube 1 as described above (see Determination of Nitrogen), but, unlike the determination of nitrogen, it is not necessary to displace air from the apparatus.

Tube 1 is heated in an electric furnace or with gas burners, as described above. After heating is completed, the cooled tube is wrapped in a filter paper and crushed, and the whole is transferred to a tall, dry, 500-ml beaker. The contents of the beaker are treated with 10-70 ml of 5% HNO_3, which is added dropwise at first, to prevent spattering of the solution. Acid is added until the calcium is completely dissolved. The solid residue (carbon, glass) is filtered off, and the filter with the solid residue is washed carefully with distilled water. 20 ml of 0.05 N $AgNO_3$ is added to the acid filtrate, which is then titrated with 0.05 N NH_4SCN in the presence of 0.5 ml of saturated ferric ammonium alum solution.*

The results of some determinations are given in the table.

DETERMINATION OF SULFUR AND NITROGEN IN A SINGLE SAMPLE OF AN ORGANIC COMPOUND

The weighed sample and 1-1.5 g of calcium are placed in the glass tube 1 of the apparatus shown in Figs. 1 and 2 (see Determination of Nitrogen). Air in the apparatus is displaced by ether or absorbed by calcium, as described in the determination of nitrogen.

The tube 1, prepared for analysis, is heated by the method described above. After heating is completed, the cooled tube 1 is wrapped in a filter paper, carefully crushed, and transferred to the flask of an apparatus for distillation of H_2S (see Determination of Sulfur). The reaction mass is decomposed with 80-100 ml of 4% H_2SO_4 or HCl, H_2S is distilled off, and sulfur is determined iodometrically (see Determination of

*Since commercial metallic calcium always contains a small amount of chlorine, an appropriate correction must be made for its chlorine content (blank experiments).

Sulfur). After distilling off H_2S, the condenser and Kjeldahl adapter are washed with distilled water, and the vessels for H_2S absorption are replaced by a vessel for absorption of ammonia (see Determination of Nitrogen). The residue in the distillation flask is made alkaline with 50% alkali, and after distilling off ammonia, the nitrogen content is determined.

The results of some determinations are given in the table.

DETERMINATION OF NITROGEN AND HALOGEN IN A SINGLE SAMPLE

The filling of tube 1, displacement of air from the apparatus, and heating of tube 1 are carried out by the method described above (see Determination of Nitrogen and Determination of Halogen).

The reaction mass is transferred to an ammonia-distillation flask, decomposed with water, ammonia is distilled off, and the nitrogen content of the compound is determined (see Determination of Nitrogen). After distilling off ammonia, the residue in the distillation flask is acidified with 5% HNO_3 until the calcium hydroxide is completely dissolved, carbon and glass are filtered off from the residue, the precipitate on the filter is washed with distilled water, and halogen is determined in the filtrate (see Determination of Halogen).

The results of some determinations are given in the table.

DETERMINATION OF SULFUR AND HALOGEN IN A SINGLE SAMPLE

The analysis is carried out in the apparatus shown in Fig. 1. The reaction mass is transferred to a flask for distillation of H_2S, decomposed with acid (see Determination of Sulfur), and, after distilling off H_2S, sulfur is determined by iodometric titration. Carbon and glass particles are filtered off from the residue in the flask after distilling off H_2S. The precipitate on the filter is washed with distilled water, and halogen is determined by the Volhard method (see Determination of Halogen).

The results of some determinations of sulfur and halogen are given in the table.

QUANTITATIVE DETERMINATION OF NITROGEN, SULFUR, AND HALOGEN IN A SINGLE SAMPLE OF AN ORGANIC COMPOUND

The decomposition of the organic compound is carried out in the apparatus shown in Fig. 1 or 2. Introduction of the sample, filling of tube 1, and removal of air are carried out by the method described for the determination of nitrogen. After heating is completed, the cooled tube is crushed in a filter paper and transferred to the dry flask of an apparatus for distillation of H_2S, the reaction mass is decomposed with acid, H_2S is distilled off, and sulfur is determined (see Determination of Sulfur). To determine nitrogen, the vessels for H_2S absorption are replaced by vessels for absorption of ammonia (see Simultaneous Determination of Sulfur and Nitrogen). The acid solution in the distillation flask is made alkaline and, after distilling off ammonia, nitrogen is determined.

Compound	%N Calc.	%N Found	%N Δ	%S Calc.	%S Found	%S Δ	%X Calc.	%X Found	%X Δ
p-Aminobenzoic acid, $C_7H_7NO_2$	10. 21	10. 13	-0. 08						
Azobenzene, $C_{12}H_{10}N_2$	15. 37	15. 36	-0. 01						
1-Naphthylamine, $C_{10}H_9N$	9. 78	9. 80	+0. 02						
Diphenylamine, $C_{12}H_{11}N$	8. 27	8. 29	+0. 02						
Nitrocinnamic acid, $C_9H_7NO_4$	7. 25	7. 01	-0. 24						
Dimethylaminobenzene, $C_{14}H_{15}N_3$	18. 66	18. 68	+0. 02						
Acridine, $C_{13}H_9N$	7. 81	7. 68	-0. 13						
Sulfosalicylic acid, $C_7H_{10}O_8S$(sic)				12. 61	12. 51	-0. 10			
m-Xylyltaurine, $C_{10}H_{15}NO_3S$				13. 98	13. 81	-0. 17			
Sodium 3-methyl-5-(2-furyl) pyrazoline-1-dithiocarboxylate $C_9H_9N_2NaOS_2$				25. 82	25. 78	-0. 04			
Hexachloroethane, C_2Cl_6							89. 89	89. 87	-0. 02
Dibromobenzene, $C_6H_4Br_2$							67. 70	67. 53	-0. 17
1-Phenyl-2-carbethoxy-4-chloro-5-hydroxybenzofuran, $C_{17}H_{13}ClO_4$							11. 19	11. 23	+0. 04
1-Phenyl-3-carbethoxy-7-chloro-5-hydroxybenzofuran, $C_{17}H_{13}ClO_4$							11. 19	10. 94	-0. 25
1-Chloro-2, 3-dimethyl-4-phenyl-2-butene, $C_{12}H_{15}Cl$							18. 21	18. 34	+0. 13
1-Chloro-4-(*p*-chlorophenyl)-2-butene $C_{10}H_{10}Cl_2$							35. 26	35. 07	-0. 19

p-Toluenesulfonamide, $C_7H_9NO_2S$	8.18	8.01	-0.17	18.72	18.56	-0.16			
Diphenylthiourea, $C_{13}H_{12}N_2S$	12.27	12.08	-0.19	14.04	13.83	-0.21			
2-Mercaptobenzothiazole, C_7H_5NS	8.35	8.26	-0.09	38.31	38.08	-0.23			
3, 5-Diphenylpyrazoline-1-thiocarboxanilide $C_{22}H_{19}N_3S$	11.76	11.49	-0.27	9.00	9.03	+0.03			
S-(2-cyanoethyl) 5-(2-furyl)-3-methylpyrazoline-1-dithiocarboxylate, $C_{12}H_{13}N_3OS_2$	15.04	15.32	+0.28	22.95	23.00	+0.05			
3-Methyl-5, 5-pentamethylenepyrazoline-1-thiocarbanilide (sic), $C_{16}H_{21}N_3S$	14.62	14.44	-0.18	11.15	10.83	-0.32			
1, 3-Dimethyl-5-(1-amino-2-sulfoethyl)-benzene, $C_{10}H_{15}NO_3S$	6.11	6.36	+0.25	13.98	14.08	+0.10			
1-Phenyl-2-methyl-3-acetyl-5-(*p*-bromobenzyloxy) indole, $C_{24}H_{18}BrNO_3$	3.12	3.14	+0.02				17.82	17.87	+0.05
1-Ethyl-2-methyl-3-carbethoxy-4,7-dichloro-5-methoxyindole, $C_{15}H_{17}Cl_2NO_3$	4.24	4.40	+0.16				21.47	21.19	-0.28

After removing sulfur and nitrogen from the flask, the contents are acidified with HNO_3 until the calcium hydroxide is completely dissolved. The solid residue is filtered off and carefully washed on the filter with distilled water; halogen in the filtrate is determined argentometrically (see Determination of Halogen).

Some results of this series of analyses are given in the table.

SUMMARY

A method is described for the quantitative determination of nitrogen, sulfur, and halogen in organic compounds by heating them with powdered calcium.

The method makes possible the quantitative determination of one, two, or three of the above elements in a single sample of an organic compound. The use of calcium speeds up and simplifies the analysis and prevents the formation of explosive mixtures.

LITERATURE CITED

[1] P. N. Fedoseev and N. P. Ivashova, *Zhur. Anal. Khim.* 7, 112 (1952); 11, 233 (1956). W. Schöniger, *Mikrochim. Acta* 1954, 74; 1955, 44.

5

The Analysis of Used Si-Cu Alloys By "Wet Oxidation" with Chromsulfuric Acid *

A. P. Terent'ev, B. M. Luskina,
and
S. V. Syavtsillo

A silicon-copper alloy, consisting of about 20% Cu and 80% Si, is used for the direct synthesis of alkyl- and arylchlorosilanes. Alloys used in the reaction (spent alloys) consist of free silicon and metallic copper, and also contain carbon and metal chlorides as impurities [1]. For reuse or regeneration of these alloys, it is important to know their composition.

This paper describes the determination of C, Cl, Si, Cu, and Fe in a single sample of spent alloy by the "wet combustion" method.

The combustion is carried out (see Fig. 1) by heating the sample in flask 1 with a mixture of concentrated sulfuric acid and a concentrated aqueous solution of chromium trioxide in a stream of oxygen. The oxidation products pass into a quartz tube 2, heated to 700-750°, containing, as catalyst, chromium sesquioxide supported on pumice, 3 (small pieces of pumice are saturated with a 10% solution of ammonium bichromate, dried, and ignited). The gaseous reaction products are then passed through a trap 4 to absorb chlorine, moisture is removed in absorbers 5 and 6, and an Ascarite tube 7 serves to absorb CO_2.

**Zavodskaya Lab.* 25, 288-289 (1959)

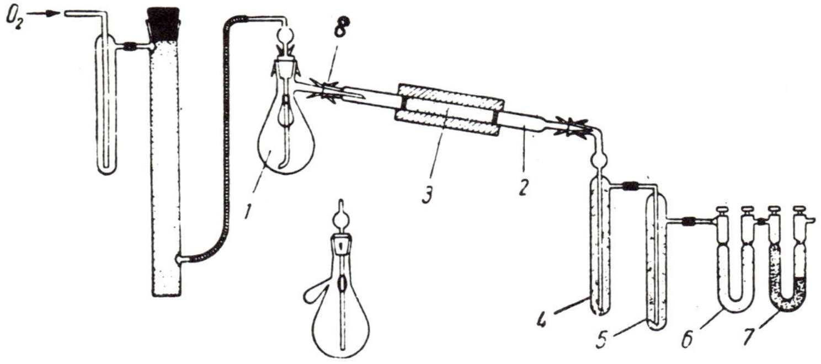

Fig. 1 Apparatus for alloy analysis by the "wet combustion" method.

Chromyl chloride is formed in flask 1, and is partly decomposed there by water vapor, liberating HCl. This compound is completely decomposed in tube 2, forming chlorine and chromium sesquioxide. To absorb chlorine and HCl, we used hydrazine hydrate, neutralized with acetic acid to pH 6. This solution does not retain CO_2. [2]

All the metals in the alloy dissolve in the oxidizing mixture during the wet oxidation and remain in the reaction flask in the form of sulfates, except free Si, which is unchanged under the conditions of the analysis and can be determined by weighing. Impurities of SiO_2, SiC, and other compounds insoluble under the conditions of the analysis are weighed together with the Si.

Fe and Cu were determined by precipitation of Cu with Na_2S, followed by solution and iodometric titration. Fe, which remains in solution as Fe^{++}, is titrated with $K_2Cr_2O_7$ in the presence of diphenylamine after removal of CuS.

ANALYTICAL PROCEDURE

Into flask 1 there is placed 10 ml conc. H_2SO_4 (heated for 1-2 hours to destroy organic impurities) and, by means of a funnel with a bent tube, 3-4 ml chromic acid (70g CrO_3 in 100 ml H_2O) is introduced into the side bulb. The flask is connected to tube 2, and a stream of oxygen is passed through at 50-60 ml/min throughout the combustion (30 minutes). The stopper is raised, and a weighing bottle with 0.2-1.5g of alloy is quickly dropped into the flask. The flask is rotated around the ground joint 8, lubricated with phosphoric acid, and 2/3 of the volume of chromic acid is added dropwise from the side bulb. The remainder is added as the red-brown vapors of chromyl chloride are displaced. The flask is then placed into a bath heated to 150-160° (preferably containing organosilicon liquid No. 5) so that the lower part of the side bulb touches the surface of the heated liquid. With this heating method, water vapor from the side bulb condenses on the walls of the flask and decomposes deposited chromyl chloride. Carbon is determined by weighing the CO_2 in tube 7; Cl in the contents of vessel 4 is determined by the Volhard method.

After determining C and Cl, the reaction flask is disconnected, methanol or ethanol and 40-50 ml water are carefully added dropwise to its contents, and it is heated for another 10-15 minutes. The precipitate is filtered through a weighed glass filter with a disc of No. 2 porosity, washed with water and a small amount of alcohol, and dried at 110-130° to constant weight. The filtrate and wash water, after removal of Si, are evaporated to approximately 100 ml, cooled to room temperature, and 5-7 ml of 10% Na_2S solution is added in portions until a persistent odor of H_2S appears. The precipitated CuS is filtered off through a glass filter

Results of the Analysis of Alloys (%)

Samples	C	Cl	Si	Cu	Fe
Alloy after synthesis of methylchlorosilanes No. 1	0.4	2.8	84.3	9.6	2.7
	0.3	2.9	84.4	9.5	2.5
Alloy after synthesis of methylchlorosilanes No. 2.	7.7	0.8	60.1	29.5	2.4
	8.0	0.8	60.4	29.0	2.8
Alloy after synthesis of methylchlorosilanes No. 3.	3.2	0.0	83.5	12.2	0.9
	3.0	0.0	84.0	12.0	1.0
Alloy after synthesis of methylchlorosilanes No. 4.	1.0	0.9	57.0	40.4	0.6
	1.1	0.7	57.0	40.5	1.0
Alloy after synthesis of phenylchlorosilanes No. 1	33.9	8.8	34.5	17.8	0.0
	33.5	8.3	34.8	18.0	0.0
Alloy after synthesis of phenylchlorosilanes No. 2	2.1	0.0	58.5	36.5	0.0
	2.1	0.0	58.3	36.7	0.0

with a disc of No. 3 or No. 4 porosity, washed, dissolved in hot HNO_3, 5 ml H_2SO_4 added, and evaporated until dense white fumes are evolved. The residue is diluted with water, 20 ml of 15% KI solution is added, and the liberated iodine is titrated with 0.1 N $Na_2S_2O_3$.

The filtrate after removal of CuS is transferred to a 250-ml volumetric flask and made up to the mark. 5 ml H_2SO_4 is added to 50 ml of the solution obtained, and the mixture is heated until the odor of H_2S disappears. 5 ml H_3PO_4 and 2-3 drops of a sulfuric acid solution of diphenylamine are added to the cooled solution, and the reduced iron is titrated with 0.1 N $K_2Cr_2O_7$ until a blue color appears.

The analysis takes 2.5-3 hours; the results of the analysis of samples of spent alloys are shown in the table.

LITERATURE CITED

[1] K. A. Andrianov, S. A. Golubtsov, et al., *Dokl. Akad. Nauk S.S.S.R.* **108**, 456 (1956); *Zhur. Priklad. Khim.* **30**, 1277 (1957).

[2] A. P. Terent'ev and B. M. Luskina, *Zhur. Anal. Khim.* **14**, 112 (1959).

6

*The Determination of Carbon Impurities in Si-Cu Alloys**

A. P. Terent'ev, B. M. Luskina,
and
S. V. Syavtsillo

A silicon-copper alloy is used in the direct synthesis of alkyl- and arylchlorosilanes. After use in the reaction, the spent alloy contains sometimes up to 20% carbon and small amounts of chlorosilanes. The carbon content in such an alloy is one of the factors determining the degree of its recovery.

We have described previously a method for determining carbon, chlorine, silicon, copper, and iron in a single sample of a spent silicon-copper alloy. [1] To determine carbon only, a simpler apparatus can be used (see Fig. 1), in which the quartz tube with ground joints is replaced by a porcelain tube.

The combustion is carried out by heating a weighed alloy sample with concentrated sulfuric acid and a concentrated aqueous solution of chromium trioxide in a stream of oxygen. The oxidation products then enter a porcelain tube containing chromium sesquioxide, heated to 700°. The gaseous reaction products are scrubbed to remove chlorine and moisture, and are passed through an Ascarite tube to absorb the CO_2 formed. Chromium sesquioxide as a supplementary

---

Plasticheskie Massy* **1960, *No*. 5, 65-66.

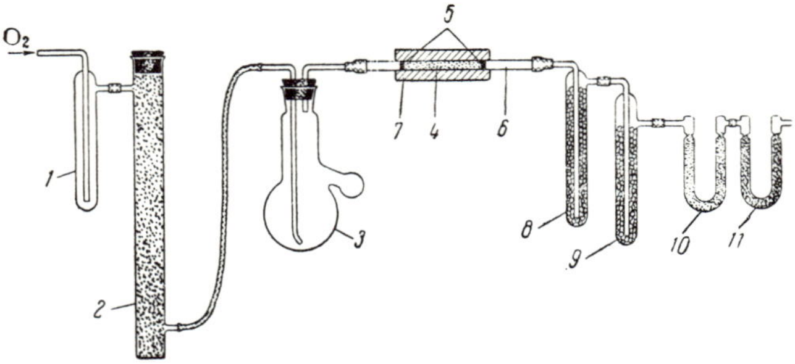

Fig. 1 Apparatus for analysis. 1: Vessel with sulfuric acid; 2: Ascarite column; 3: flask; 4: catalyst; 5: asbestos plugs; 6: heater; 7: porcelain tube; 8-11: absorption tubes.

combustion catalyst is particularly suitable for the determination of carbon in the presence of chlorine-containing substances, as in our case.

In the presence of HCl, formed in the hydrolysis of chlorosilanes, the formation of volatile, red-brown chromyl chloride is observed. Its vapors, on passing through the heated chromium sesquioxide tube, decompose according to the equation:

$$4\ CrO_2Cl_2 \rightarrow 2\ Cr_2O_3 + Cl_2 + O_2$$

Thus, the catalyst used by us not only is not poisoned, but is continuously regenerated as a result of the formation of Cr_2O_3. Partially condensed chromyl chloride on the walls of the reaction flask is decomposed by water vapor, forming HCl.

To absorb the chlorine and HCl formed as a result of the decomposition of chromyl chloride by the method described, we have used a solution of hydrazine hydrate neutralized to pH 6 with acetic acid. [2] This quantitatively absorbed halogen and did not retain CO_2. Dilute HCl may also be used to neutralize hydrazine hydrate in this case. [3]

ANALYTICAL PROCEDURE

A 0. 1-1.5 g sample of alloy (depending on the carbon content) is placed in flask 3, whose ground joint is lubricated with phosphoric acid, 10 ml of concentrated H_2SO_4 (heated for 1-2 hours to destroy organic impurities) is added, and 2-3 ml of chromic acid (70g CrO_3 dissolved in 100 ml H_2O) is placed in the side bulb by means of a funnel with a bent tube. The flask and the weighed absorption tube 11 are connected as shown in Fig. 1. By tilting the flask, the chromic acid solution is poured from the side bulb into the lower part of the flask, which is heated at 150-160° in a bath containing an organosilicon liquid.

Oxygen passing through vessel 1 containing concentrated H_2SO_4 enters the Ascarite column 2, which removes CO_2 impurities. The stream of purified oxygen, maintained at a flow rate of 50-60 ml/min throughout the analysis, and the reaction products from the flask pass into the heated (700-750°) porlain tube 6, containing Cr_2O_3 supported on pumice 4. The oxidation products then pass successively through the absorption tubes 8 (containing hydrazine hydrate, neutralized accurately to pH 6 with acetic acid), 9 (concentrated H_2SO_4), and 10 (Anhydrone). Tube 11, filled with Ascarite [2, 3] and Anhydrone [1, 3] absorbs the CO_2 formed in the combustion of the sample.

It was found in practice that 2-3 combustions can be carried out in the flask without disconnecting it and without changing the H_2SO_4 in it, by merely introducing 2-3 ml of chromic acid into the side bulb for each new sample.

The method does not require continuous supervision; two sets of apparatus can be operated simultaneously. The time required for one combustion is 30 minutes.*

We carried out parallel determinations of carbon in spent alloys by the method proposed and by the rapid combustion method. [4] The results of these determinations are shown in Table 1.

Table 1 shows the satisfactory agreement between the data of parallel determinations by different methods. It should be noted that carbon bonded to silicon as silicon carbide is not determined by the wet combustion method.

A considerable number of spent alloys were analyzed by the proposed method. Some results of these determinations are given in Table 2.

---

* The catalyst (Cr_2O_3 supported on pumice) is prepared as follows: Pumice, ground to a grain size of 2-3 mm and screened to remove dust, is washed first with hot HNO_3, then treated with HCl, until a colorless filtrate is obtained. The pumice is finally washed with hot distilled water until a negative reaction is obtained for chloride ion, and is dried at 120°. The pure pumice obtained is impregnated with 10% ammonium bichromate solution, dried, and ignited for 1-1.5 hours. The finished catalyst is placed, as described above, in the porcelain tube, between two loose asbestos plugs 5. The catalyst thus prepared has mechanical strength, and its catalytic properties do not change appreciably with prolonged use.

TABLE 1

Results of the Determination of Carbon in Spent Alloys by Different Methods.

Sample No.	%C By wet combustion	%C By rapid combustion	Deviation
1	13.1	13.1	0.0
	12.9	12.7	0.2
2	2.0	2.2	0.2
	2.2	2.4	0.2
3	2.0	2.2	0.2
	1.9	2.0	0.1

TABLE 2

Results of the Determination of Carbon in Spent Alloys.

Sample No.	g Sample	g CO_2	%C	Deviation between Parallel Determinations.
4	0.2132	0.1454	18.6	0.0
	0.2192	0.1496	18.6	
5	0.4341	0.1220	7.7	0.1
	0.4368	0.1212	7.6	
6	1.5291	0.1222	2.2	0.1
	1.1245	0.0929	2.3	
7	1.6098	0.1605	2.7	0.3
	1.1121	0.1240	3.0	

SUMMARY

A simplified method for the determination of carbon in spent copper-silicon alloys has been developed. The time required for one combustion is 30 minutes.

LITERATURE CITED

[1] A. P. Terent'ev, B. M. Luskina, and S. V. Syavtsillo, *Zavodskaya Lab.* 25, 288 (1959).
[2] A. P. Terent'ev and B. M. Luskina, *Zhur. Anal. Khim.* 14, 112 (1959).
[3] A. A. Houghton, *Analyst* 70, 118 (1945).
[4] M. O. Korshun and N. E. Gel'man: "New Methods of Elemental Microanalysis", Moscow-Leningrad (1949).

7

*Elemental Organic Analysis by "Wet Oxidation"**

A. P. Terent'ev and B. M. Luskina

The analysis of organic compounds by complete oxidation, using oxidizing agents in solution, is called the wet combustion method. An example is the Kjeldahl nitrogen determination. Similar methods have been known for a long time and are frequently used for the decomposition of organic compounds to determine one element. They are rarely used for the simultaneous determination of several elements. The first to indicate this possibility in the analysis of organometallic compounds was Messinger [1] in 1888.

The most frequently used oxidants are chromic acid or its salts, mixed with concentrated sulfuric acid, or with sulfuric and phosphoric acids. Literature reviews have been given by Houghton [2] and Van Slyke [3].

As a rule, it is necessary to carry out a "complete combustion"; for this, the gaseous reaction products and oxygen are generally passed through a heated tube containing copper oxide. However, this method has a number of significant disadvantages. Specifically, great difficulties arise in the determination of halogen-containing compounds [4].

In our proposed method, chromium sesquioxide was used as catalyst in the complete combustion. This

Zhur. Anal. Khim.* **14, 112-117 (1959)

enabled us to avoid a number of difficulties.

Chromium sesquioxide has been used as a surface-combustion catalyst by Fedoseev and coworkers [5] in elemental analysis. Klimova and coworkers [6] used it in the microelemental analysis of organosilicon compounds.

The combined combustion method that has been developed, which may be called the chromic method, is of great interest, particularly for industrial laboratories. The direct determination of hydrogen, sulfur, and chromium is obviously impossible, but the method does permit determination of many other elements, together with carbon and nitrogen, in one sample. The analysis can be carried out not only with dry samples, but also with solutions of organic compounds in water or sulfuric acid. The simplicity of the apparatus and the possibility of working with a wide range of samples make the method very convenient.

Determination of Carbon

The oxidation mixture used by us consists of concentrated sulfuric acid and a concentrated aqueous solution of chromic acid, which is added gradually, in small portions, during the oxidation. To ensure complete combustion, especially prepared chromium sesquioxide, deposited on small pieces of pumice, was used. Combustion of the organic samples was carried out in the apparatus shown in Fig. 1.

10 ml of concentrated sulfuric acid is placed into the reaction flask 3, and 1-2 ml of chromic acid into its side bulb. A stream of oxygen is then passed through at a rate of about 50-60 ml/min, the weighed tube 11 is connected, the stopper is raised, and a capsule with the sample is dropped into the reaction

flask. Chromic acid is gradually added dropwise from the side bulb by carefully rotating the flask around the ground joint b which, like the ground joints a and c, is lubricated with phosphoric acid. The flask is immersed in a bath heated to 150-160°. The gaseous reaction products, which may also contain products of incomplete oxidation (CO, CH_3COOH, etc.), together with a stream of oxygen, pass into a heated (700-750°) tube containing chromium sesquioxide, where they undergo complete oxidation; the oxidation products then pass through vessel 8 (whose purpose is indicated below), concentrated acid 9, and Anhydrone 10, to purify and dry the CO_2 formed. Carbon is determined from the increase in weight of tube 11, filled with Ascarite and Anhydrone. Tube 12 (not shown in the figure), with the same filling, is connected to tube 11. The combustion time does not exceed 30 minutes.

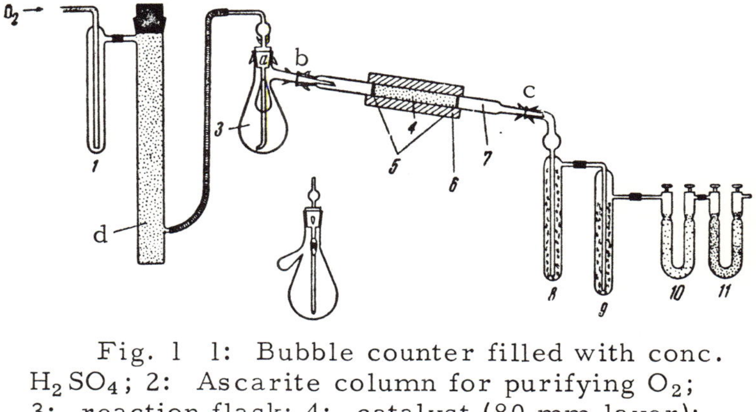

Fig. 1 1: Bubble counter filled with conc. H_2SO_4; 2: Ascarite column for purifying O_2; 3: reaction flask; 4: catalyst (80 mm layer); 5: loose asbestos stoppers; 6: electric furnace (700-750°); 7: quartz tube (length 200 mm, diameter 10-12 mm); 8: vessel containing glass rods (in the bulb, a plug of glass wool) moistened with an acetic acid solution of hydrazine hydrate; 9: vessel containing glass rods moistened with conc. H_2SO_4; 10: Ascarite tube; 11: tube with Ascarite and Anhydrone.

The simultaneous determination of carbon and other elements was carried out in the same apparatus. Changes in the method were introduced only in the case of combustion of chlorine-containing compounds.

Simultaneous Determination of Carbon and Chlorine (or Bromine)

In the oxidation of chlorine-containing organic compounds by the chromic method, volatile chromyl chloride is formed, which decomposes in the chromium sesquioxide tube, liberating chlorine and oxygen and forming chromium sesquioxide. Chromyl chloride is also partially decomposed by water vapors, liberating hydrogen chloride. The copper oxide used earlier forms volatile chlorine compounds in this case. Bromine-containing organic compounds are oxidized with liberation of elemental bromine.

A solution of hydrazine hydrate, neutralized with acetic acid to pH 6, was used to absorb the halogens. Under these conditions, the halogen is completely absorbed and the CO_2 is not retained.

The sample of halogen-containing compound is burned as described for the determination of carbon. In the case of chlorine-containing compounds, the amount of chromic acid used is 1-2 ml greater, and it is gradually added dropwise to the lower part of the flask before heating is started, as the red-brown vapors of chromyl chloride are displaced from the flask; about one third of the original volume remains in the side bulb. The flask 3 is then lowered into the heated bath so that the side bulb is in contact with the liquid in it. This leads to decomposition of the chromyl chloride deposited on the walls of the flask by water vapor. The halogens are absorbed in vessel 8, containing glass rods moistened with an acetic acid

TABLE 1. Determination of Carbon

Compound	mg sample	mg CO_2	% C Found	% C Calc.	% C Δ
Salicylic acid, $C_7H_6O_3$	110.80	245.99	60.59	60.87	-0.28
	70.94	158.24	60.88		+0.01
Ethylanthraquinone, $C_{16}H_{12}O_2$	60.81	181.54	81.48	81.34	+0.14
	59.84	178.63	81.47		+0.13
Acenaphthene, $C_{12}H_{10}$	54.43	186.97	93.75	93.46	+0.29
	54.22	186.03	93.63		+0.17
Cholesterol, $C_{27}H_{46}O$	53.68	164.68	83.72	83.87	-0.15
	60.12	184.74	83.86		-0.01
Xanthone, $C_{13}H_8O_2$	49.82	145.02	79.44	79.58	-0.14
	52.38	152.83	79.63		+0.05
Diethyl dimethylbenzodi-furandicarboxylate, $C_{18}H_{18}O_6$	64.84	155.64	65.51	65.45	+0.06
	60.59	145.68	65.62		+0.17

TABLE 2. Simultaneous Determination of Carbon and Chlorine (or Bromine).

Compound	% C			%X		
	Found	Calc.	Δ	Found	Calc.	Δ
p-Chlorobenzoic acid $C_7H_5ClO_2$	53.50	53.70	-0.20	22.44	22.65	-0.21
	53.57		-0.13	22.54		-0.11
2, 6-Dichloroquinone $C_6H_2Cl_2O_2$	40.61	40.72	-0.11	39.96	40.06	-0.10
	40.82		+0.10	40.10		+0.04
1-Chloro-4-nitrobenzene $C_6H_4ClNO_2$	45.83	45.74	+0.09	22.44	22.50	-0.06
	45.62		-0.12	22.35		-0.15
4-Chloro-6-isopropyl-*m*-cresol $C_{10}H_{13}ClO$	64.93	65.04	-0.11	19.35	19.20	+0.15
	65.12		+0.08	19.25		+0.05
Chlorodinitrobenzene $C_6H_3ClN_2O_4$	35.38	35.58	-0.20	17.34	17.50	-0.16
	35.67		+0.09	17.56		+0.06
2, 3-Dibromo-1, 4-naphthoquinone $C_{10}H_4Br_2O_2$	38.24	38.01	+0.23	50.51	50.58	-0.07
	38.32		-0.31	50.31		-0.27
2-Bromo-1, 4-naphthoquinoneaceto-acetic ester, $C_6H_4BrNO_2$	52.74	52.62	-0.12	22.06	21.88	-0.18
	52.69		+0.07	21.83		-0.05
Bromonitrobenzene $C_6H_4BrNO_2$	35.59	35.67	-0.08	39.40	39.56	-0.16
	35.79		+0.12	39.60		+0.04

solution of hydrazine hydrate (pH 6). The remaining operations are carried out as in the carbon determination. The chloride or bromide ions in the contents of vessel 8 are titrated by the Volhard method.

Thus, the use of chromium sesquioxide as catalyst for complete combustion, and of an acetic acid solution of hydrazine hydrate to absorb halogens, has made it possible for us to determine carbon and chlorine (or bromine) simultaneously.

In the analysis of compounds for carbon, or for carbon and halogen, it is possible to burn two or even three samples in the same flask. The flask is not disconnected, and the sulfuric acid is not changed; one merely introduces a new sample and more chromic acid.

Simultaneous Determination of Carbon, Chlorine, and Nitrogen

If nitrogen-containing compounds give ammonia in concentrated sulfuric acid, then nitrogen can also be determined together with carbon and chlorine. The contents of reaction flask 3 are transferred to a Kjeldahl flask after combustion, excess alkali is added, and ammonia is distilled in the usual apparatus into a standard acid solution. Nitrogen in amines, nitriles, pyrrolidines, indoles, etc., can be determined in this manner.

It is interesting to note that nitrogen can be determined in these compounds in the presence of other forms of nitrogen ($-NO_2$, -NH-NH-, etc.). Examples of such determinations are given in Table 4.

Thus, in a number of cases, the wet combustion method can be used for the separate determination of different forms of nitrogen.

The method of simultaneous determination of

TABLE 3. Simultaneous Determination of Carbon, Chlorine, and Nitrogen

Compound	% C Found	% C Calc.	% C Δ	% Cl Found	% Cl Calc.	% Cl Δ	% N Found	% N Calc.	% N Δ
Compound of salicylic acid and N-ethyl-α-methylpyrrolidine hydrochloride, $C_{14}H_{20}ClNO_3$	58.61 58.87	58.85	-0.24 +0.02	12.18 12.25	12.41	-0.23 -0.16	4.90 5.06	4.90	0.0 +0.16
Compound of cinnamic acid and N-ethyl-α-methylpyrrolidine hydrochloride, $C_{16}H_{22}ClNO_2$	65.00 64.88	64.96	+0.04 -0.08	11.87 11.82	11.93	-0.11 -0.16	4.69 4.83	4.74	-0.05 +0.09
Naphthyl compound of N-propyl-α-methylpyrrolidine hydrochloride, $C_{18}H_{24}ClNO_2$	67.26 67.11	67.17	+0.09 -0.06	11.25 11.20	11.02	+0.23 +0.18	4.52 4.49	4.38	+0.17 +0.14
Ethyl oxamate, $C_4H_7NO_3$	41.30 41.03	41.02	+0.28 +0.01				12.25 11.99	11.96	+0.29 +0.03
Succinonitrile, $C_4H_2N_2$	61.46 61.78	61.53	-0.07 +0.25				35.84 36.01	35.88	-0.04 +0.13
5-Acetoxy-3-carbethoxy-2-methyl-1-phenylindole, $C_{20}H_{19}NO_4$	71.42 71.49	71.20	+0.22 +0.29				4.00 4.25	4.15	-0.15 +0.10

TABLE 4. Simultaneous Determination of Carbon and Nitrogen in Compounds Containing Two Forms of Nitrogen ("Oxidized" and "Reduced").

Compound	%C			%N*		
	Found	Calc.	Δ	Found	Calc.	Δ
4-(2-Cyanoethyl)-3-methyl-5-pyrazolone, $C_7H_9N_3O$	55.62	55.62	0.0	9.41	9.26	+0.15
	55.61		-0.01	9.44		+0.18
1-(2-Cyanoethyl)-3-isobutyl-5-pyrazolone, $C_{10}H_{15}NO$	62.42	62.15	+0.27	7.30	7.25	+0.05
	62.24		+0.09	7.22		-0.03
4-(2-Cyanoethyl)-3-methyl-1-phenyl-5-pyrazolone, $C_{13}H_{13}N_3O$	68.67	68.70	-0.03	6.30	6.16	+0.14
	68.78		+0.08	6.15		-0.01
Tris(2-cyanoethyl) ether of tris-(hydroxymethyl) nitromethane $C_{13}H_{18}N_4O_5$	50.55	50.32	+0.23	13.65	13.54	+0.11
	50.49		+0.17	13.76		+0.22
m-Nitroaniline, $C_6H_6N_2O_2$	52.22	52.17	+0.05	10.18	10.14	+0.06
	52.19		+0.02	10.08		-0.04

*Nitrogen content was calculated for the -CN and $-NH_2$ groups.

carbon, halogen, and nitrogen is also suitable for the analysis of compounds containing other elements. The combustion method remains the same. In the absence of halogens, vessel 8 can be omitted.

Analysis of Organic Compounds Containing Metals

In the oxidation of organic compounds containing metals, the latter generally remain in the reaction flask in the form of sulfates, since complete mineralization occurs in this case. The presence of chromic acid may cause difficulties in the determination of metals. At the end of the combustion, Cr(VI) is therefore reduced to Cr(III) by the addition of a few drops of methanol. Complex compounds containing iron or copper were used as test samples.

To determine iron, the sulfate solution was treated with sodium sulfide, and, after removing H_2S, the reduced iron was titrated with potassium dichromate in the presence of diphenylamine.

In the analysis of copper complexes, copper was determined iodometrically in the sulfate solution. This solution was used further for the determination of nitrogen, by distilling the ammonia formed in the usual way. Table 5 shows some examples of the analysis of complex compounds containing iron or copper.

Analysis of Organosilicon Compounds

In the combustion of these compounds, silicon is completely oxidized and remains in the form of a gel of silicic acid. A 2% gelatin solution was added to precipitate and accelerate the filtration of silicic acid from the sulfate solution. Filtration, ignition, and weighing were carried out in porcelain crucibles with

TABLE 5. Simultaneous Determination of Carbon and Iron, or Carbon, Nitrogen, and Copper.

Compound	% C Found	% C Calc.	% C Δ	% N Found	% N Calc.	% N Δ	% M Found	% M Calc.	% M Δ
Triisovaleryliron(III) triacetonate $C_{24}H_{39}FeO_6$	60.06 60.00	60.13	-0.07 -0.13				11.45 11.35	11.65	-0.20 -0.30
Ferrocene, $C_{10}H_{10}Fe$	64.77 64.73	64.56	+0.21 +0.17				30.16 29.93	30.02	+0.14 -0.09
Diacetylferrocene, $C_{14}H_{14}FeO_2$	62.43 62.42	62.25	+0.18 +0.17				20.56 20.77	20.68	-0.12 +0.09
Copper methylaminoacetylacetonate $C_{12}H_{20}CuN_2O_2$	50.14 50.24	50.06	+0.08 +0.18	9.70 9.66	9.73	-0.03 -0.07	22.06 22.29	22.07	-0.01 +0.22
Copper(II) bis(α-ethylamino-o-cresol) $C_{18}H_{20}CuN_2O_2$	60.10 60.35	60.07	+0.03 +0.28	7.74 7.62	7.78	-0.04 -0.16	17.44 17.54	17.65	-0.21 -0.11
Copper(II) bis(α-methylamino-o-cresol) $C_{16}H_{16}CuN_2O_2$	57.79	57.91	-0.12	8.47	8.44	+0.03	19.04	19.15	-0.11

porous bottoms. When paper filters were used, appreciable absorption of Cr(III) salts was observed. In this case, it is necessary to wash the filter and precipitate with hot hydrochloric acid (1:1).

Examples of the analysis of some organosilicon compounds are shown in Table 6.

Since the majority of chlorine-containing organosilicon compounds hydrolyze readily, samples of these liquid compounds are generally taken in glass ampoules. This method is obviously inapplicable for the gravimetric determination of silicon. We used the following method for taking samples of readily hydrolyzed compounds: A few drops of the compound to be analyzed are introduced, in an atmosphere of dry nitrogen, into a weighed capsule containing a small amount of anhydrous sodium sulfate, the capsule is tapped gently so that the salt covers the sample, and then reweighed in a closed weighing bottle. The sample was also covered with sodium sulfate in the analysis of solid, hygroscopic materials, or of materials that are decomposed by concentrated sulfuric acid with evolution of gaseous products (e.g., hydrochlorides). The sodium sulfate acts only as a "plug". The salt dissolves during the analysis and does not interfere with the subsequent determination of silicon or other elements.

The following results were obtained in the simultaneous determination of carbon, chlorine, and silicon in trichloro(chlorophenyl)silane:

Calc. for	% C	% Cl	% Si
$C_6H_4Cl_4Si$	29.29	57.65	11.42
Found	29.05	57.38	11.13
	29.20	58.46	11.23

TABLE 6. Simultaneous Determination of Carbon and Silicon.

Compound	% C			% Si		
	Found	Calc.	Δ	Found	Calc.	Δ
Octaethylcyclotetrasiloxane $C_{16}H_{40}O_4Si_4$	47.08	47.00	+0.08	27.43	27.48	-0.05
	47.22		+0.22	27.69		+0.21
Octamethylcyclotetra-siloxane $C_8H_{24}O_4Si_4$	32.17	32.39	-0.22	38.03	37.88	+0.15
	32.28		-0.11	37.94		+0.06
Tetrabutylsilane* $C_{16}H_{36}Si$	75.03	74.91	+0.12	10.73	10.95	-0.22
	74.74		-0.17	10.85		-0.10

*Tetraethylsilane and chromic mixture explode in an atmosphere of pure oxygen. [7]

SUMMARY

A "wet combustion" method has been developed for the simultaneous determination of carbon, halogen, and nitrogen in organic compounds. A sample of the compound is heated at 150-160° with a mixture of concentrated sulfuric acid and a concentrated aqueous solution of chromic acid. The oxidation products are carried by a stream of oxygen through a heated (700-750°) quartz tube containing a catalyst: chromium sesquioxide on pumice. Carbon is oxidized to CO_2, which is absorbed by Ascarite and weighed. Halogen is liberated in the elemental form or as hydrogen halide, and is absorbed by an acetic acid solution of hydrazine hydrate and titrated argentimetrically by the Volhard method. Nitrogen is determined after combustion by distillation of ammonia from an alkaline solution.

Iron, copper, silicon, and other elements that mineralize during the oxidation and remain in the sulfate solution can be determined simultaneously with carbon, halogen, and nitrogen.

LITERATURE CITED

[1] J. Messinger, *Chem. Ber.* 21, 2910 (1888).
[2] A. A. Houghton, *Analyst* 70, 118 (1945).
[3] D. D. Van Slyke, *Anal. Chem.* 26, 1706 (1954).
[4] W. P. Pickhardt, A. N. Oemler, and J. Mitchell, Jr., *ibid.* 27, 1784 (1955).
[5] P. N. Fedoseev and M. M. Pavlenko, *Zhur. Anal. Khim.* 8, 158 (1953); P. N. Fedoseev and M. Ya. Sobko, *ibid.* 10, 323 (1955).
[6] V. A. Klimova, M. O. Korshun, and E. G. Bereznitskaya, *Zhur. Anal. Khim.* 11, 223 (1956); V. A. Klimova and E. G. Bereznitskaya,

ibid. **11**, 292 (1956).

[7] K. A. Andrianov: "Organosilicon Compounds", Moscow: Goskhimizdat (1955), p. 100.

8

*The Rapid Determination of Silicon in Organosilicon Compounds**

A. P. Terent'ev, S. V. Syavtsillo, and B. M. Luskina

Wet oxidation methods have been used for the analysis of organosilicon compounds for a long time. [2] Kreshkov and Nessonova [3] have shown that the conditions and possibilities of oxidation by the wet method are specific for each organic compound. Therefore, they used KIO_3 in H_2SO_4 medium for the oxidation of esters of orthosilicic acid, but for diethyl- and diphenyldichlorosilane they used a more complex mixture, consisting of CrO_3 and KIO_3 in a medium of concentrated H_2SO_4 and H_3PO_4. The authors determined the SiO_2 formed in all cases gravimetrically or colorimetrically, in the form of blue silicomolybdate.

The determination of Si in polysiloxanes is presently carried out by heating with a mixture of 25% oleum and concentrated HNO_3, containing 20% nitrogen oxides. The operation is performed very carefully, to prevent foaming and spattering of the sample. After removing the nitrogen oxides and H_2SO_4, the quartz flask in which the oxidation is carried out is ignited to constant weight at 800°, and the Si content is obtained from the weight of SiO_2. This method is characterized by simplicity of execution and high

* *Zhur. Anal. Khim.* **16**, 83-86 (1961)

precision (0.2%), but is lengthy (7-8 hours). [4]

Šír and Komers [5] also decomposed organosilicon compounds with a mixture of oleum and fuming nitric acid, but finished the Si determination titrimetrically, which should have speeded up the analysis. However, the need for fusing the SiO_2 formed with potassium and sodium carbonates to convert it to a soluble state increased the total time of the analysis.

We previously [1] developed a method for determining C and Si in a single sample in organosilicon compounds. Differing from the method of Kreshkov and Nessonova [3], we carried out the oxidation with concentrated H_2SO_4 and a concentrated aqueous solution of CrO_3 in a stream of oxygen at 150°. Under these conditions, even such stable compounds as polysiloxanes are completely oxidized. Complete mineralization in the latter case was achieved considerably faster (30 minutes) than by the present method. [4] No resinification or sulfonation products were formed (these are often more resistant to oxidation then the original compounds). The mineralized Si was determined gravimetrically, which required 3-3.5 hours.

Thus, a rapid method for decomposing organosilicon compounds [1] was concluded with a lengthy gravimetric determination of Si, while a lengthy decomposition [5] used a rapid titrimetric determination of Si.

Kálmán and Vágó [6] successfully used a titrimetric determination of Si in tetraethoxysilane. The method proposed by the authors was based on the hydrolysis of tetraethoxysilane in alcoholic HF solution, with subsequent titration of the K_2SiF_6 formed. The determination takes 20-30 minutes. However, this method is suitable only for the analysis of monomeric and polymeric ethoxysilanes.

In the present work, we used for the rapid determination of Si in organosilicon compounds a method previously used by us for the mineralization of various organic and elemento-organic compounds with a mixture of chromic and sulfuric acids. [1] The silicic acid formed was filtered off [3] and dissolved in concentrated alkali solution, and the Si was then determined titrimetrically by the method of Šír and Komers. [5] In the analysis of ethoxysilanes or chlorosilanes (containing no radicals), which are easily hydrolyzed by aqueous alkali to form sodium silicate, the determination of Si becomes even simpler, since it is not necessary to heat with the acid mixture.

ANALYTICAL PROCEDURE

The sample of organosilicon compound is placed in a short-necked pear-shaped 100-ml flask containing 10 ml concentrated H_2SO_4, 2-3 ml H_2CrO_4 (70 g CrO_3 dissolved in 100 ml H_2O) is added dropwise with continuous stirring, and the mixture is heated 30 minutes at 150-160°. The flask is then cooled (avoiding external cooling with water) and 30-40 ml H_2O is added. The solution obtained is boiled 5 minutes at 110-120°. If no boiling of the solution is observed at 120°, the flask is cooled, a little more water is added, and it is again heated. If, however, boiling begins below 110°, the solution is evaporated until a temperature of 110-120° is reached. When heating is complete, the solution is cooled to 60-70°, 1-1.5 ml of 2% gelatin solution is added with vigorous stirring, and the flask is left for 5-10 minutes to complete coagulation of the silicic acid. The precipitate is filtered through a loose, ashless filter (7 cm) and washed with hot distilled water.

The filter and precipitate are transferred to a polyethylene* beaker, 5-7 ml of 30% NaOH solution is added to complete solution of the precipitate, 5-6 drops of indicator** is added, the solution is neutralized with HCl (first 1:1, then 1:10 until slightly acid), the solution obtained is transferred to an Erlenmeyer flask and exactly neutralized with 0.1 N acid and alkali solutions. The total volume must not exceed 40-50 ml. The solution is then saturated with neutral KCl to prevent hydrolysis of the fluosilicate, 10 ml of neutral NH_4F solution*** and 20 ml 0.1 N HCl are added, and the excess acid is titrated with 0.1 N alkali.

The method was tested on a number of individual organosilicon compounds (Table 1).

---

* Filtration is conveniently carried out with a polyethylene funnel, and the precipitate is dissolved in alkali in it, collecting the solution in a beaker of non-siliceous material.

** The indicator is prepared by mixing two solutions: (a) 0.1% alcoholic methyl red; (b) to 100 ml of 0.1% bromocresol green in 20% alcohol there is added 0.57 ml 0.1 N HCl. 5 parts of solution (a) are mixed with 7 parts of solution (b).

*** 1% NH_4F solution is prepared in a polyethylene flask by cooling a mixture of 20 ml 2.5% NH_4OH and 12.5 ml 40% HF. The mixture obtained is diluted to 500 ml with water and, after coarse neutralization, is neutralized exactly with 0.1 N acid and alkali. The prepared solution is checked daily for neutrality as follows: 20 ml 0.1 N HCl and 10 ml NH_4F is titrated with 0.1 N alkali in the presence of 4-5 drops of indicator. If the amount of alkali used differs from that used to titrate 20 ml of acid, the NH_4F solution is again neutralized.

TABLE 1. Determination of Silicon in Organosilicon Compounds after Oxidation with a Mixture of Chromic and Sulfuric Acids.

Compound	mg Sample	ml 0.1N HCl used	% Si Found	% Si Calc.	% Si Δ
Octadecaethyloctasiloxane $C_{36}H_{90}O_7Si_8$	27.20	10.20	26.31	26.13	+0.18
	25.40	9.41	25.99		-0.14
Octaethylcyclotetrasiloxane $C_{16}H_{40}O_4Si_4$	28.45	11.20	27.62	27.48	+0.14
	26.70	10.41	27.35		-0.13
Bis(decyloxy)dimethylsilane $C_{22}H_{48}O_2Si$	90.60	9.61	7.44	7.54	-0.10
	90.40	10.00	7.76		+0.22
(Decyloxy)trimethylsilane $C_{13}H_{30}OSi$	101.20	16.96	11.76	12.19	-0.43
	98.80	16.81	11.94		-0.25
Dibutoxychlorophenylsilane $C_{14}H_{23}ClO_2Si$	46.50	6.46	9.75	9.79	-0.04
	48.00	6.81	9.95		+0.16
Butoxychloromethylphenylsilane $C_{11}H_{17}ClOSi$	43.10	7.64	12.44	12.28	+0.16
	36.90	6.26	11.90		-0.38

TABLE 2. Determination of Silicon in Technical Ethylpolysiloxanes.

Sample Identification	mg Sample	ml 0.1 N HCl used	% Si Found: By proposed method	% Si Found: By present method	% Deviation between methods
Type 5, batch 3	13.25	5.13	27.2	26.9	0.4
	22.70	8.89	27.5		
Type 4, part 2	18.60	7.64	28.8	28.4	0.6
	32.00	13.37	29.3		
Liquid 1, batch 7	21.10	6.49	21.6	22.2	0.6
	21.80	6.75	21.7		
Liquid 2, batch 13	19.70	7.22	25.7	25.6	0.1
	20.80	7.53	25.4		
Liquid 3, batch 1	20.60	7.85	26.7	27.1	0.6
	15.90	5.97	26.3		
Liquid 5, batch 44	24.10	9.31	27.1	26.9	0.4
	14.40	5.65	27.5		

TABLE 3. Determination of Silicon in Easily Hydrolyzed Silicon-Containing Compounds.

Compound	mg Sample	ml 0.1 N HCl used	% Si Found	% Si Calc.	Δ
Triethoxy-silane $C_6H_{16}O_3Si$	49.55	12.23	17.30	17.10	+0.20
	49.70	11.98	16.90		-0.20
Silicon tetrachloride $SiCl_4$	41.85	9.93	16.64	16.53	+0.11
	50.08	12.03	16.85		+0.32

Silicon was determined in technical ethylpolysiloxanes by this method. The results of the analysis are in good agreement with the data obtained by the present method. [4] (Table 2)

Table 3 shows examples of Si determination without preliminary oxidation. In this case, the sample was hydrolyzed in 15% NaOH in a polyethylene flask. The subsequent neutralization and titration were carried out as described above. It should be noted that, in compounds of this type, hydrogen joined to silicon can also be determined together with silicon in the same sample. [7]

SUMMARY

A rapid method for determining silicon in organosilicon compounds has been developed, based on their oxidation with a mixture of chromic and sulfuric acids at 150°. The silicic acid formed is filtered off, dissolved in alkali, converted to fluosilicate ion in the presence of acid; the excess acid is titrated with alkali. The determination takes 1.5 hours. The determination of silicon in compounds easily hydrolyzed by aqueous alkali to form silicic acid takes 30 minutes.

LITERATURE CITED

[1] A. P. Terent'ev and B. M. Luskina, *Zhur. Anal. Khim.* **14**, 112 (1959).

[2] A. P. Kreshkov, et al.: "Analysis of Organosilicon Compounds", Moscow: Goskhimizdat (1954).

[3] A. P. Kreshkov and G. D. Nessonova, *Zhur. Anal. Khim.* **4**, 220 (1949).

[4] A. P. Kreshkov, S. V. Syavtsillo, and V. T. Shemyatenkova, *Zavodsk. Lab.* **22**, 1425 (1956).

[5] Z. Šír and R. Komers, *Chem. Listy* **50**, 88 (1956); *Collection Czech. Chem. Commun.* **21**, 873 (1956).

[6] L. Kálmán and Á. Vágó, *Magy. Kém. Folyóirat* **64**, 123 (1958).

[7] E. A. Bondarevskaya, S. V. Syavtsillo, and A. M. Neshumova, *Tr. Nauch.-Issled. Inst.* No. 1, Moscow: Goskhimizdat (1958), p. 95.

9

*The Simultaneous Determination of Carbon and Nitrogen in Nitro Compounds**

A. P. Terent'ev and B. M. Luskina

"Reduced" and "oxidixed" forms of nitrogen are distinguished on the basis of the behavior of nitrogen-containing substances upon oxidation. The reduced form of nitrogen appears, for example, in amines, imines, etc., which form ammonia upon oxidation with sulfuric acid. Compounds with azo, hydrazo, nitro, nitroso groups, etc., which form nitrogen as the element or as oxides upon oxidation, are considered as compounds with the oxidized form of nitrogen.

Methods are known in the literature for the simultaneous determination of C and N by wet combustion. As a rule, they are suitable for compounds with nitrogen in the reduced form. [1-3]

We have shown earlier [4] that, in the combustion of amines, nitriles, indoles, and pyrrolidines with a mixture of sulfuric and chromic acids, carbon is oxidized to CO_2 and determined gravimetrically; nitrogen is converted to $(NH_4)_2SO_4$ and determined after addition of excess alkali by distillation as NH_3. It was also found that, when reduced and oxidized forms of nitrogen are present simultaneously, only the former is determined.

The wet combustion method, and the Kjeldahl method without preliminary reduction, are unsuitable for the determination of nitrogen in many organic

* *Zhur. Anal. Khim.* **16**, 462-464 (1961)

compounds. To extend the range of application of the Kjeldahl method, various methods of preliminary reduction for compounds containing nitrogen in the oxidized form have been proposed. Among the reducing agents used are glucose [5], hydriodic acid [6], thiosalicylic acid [7], chromous chloride solution [8], a hydrochloric acid solution of stannous chloride [9], zinc dust, and others.

It has been established in the present work that the most expedient agent for the preliminary reduction of the nitro group is zinc dust in an aqueous solution of sulfuric acid. In this way it was possible to determine carbon and nitrogen in the same sample.

ANALYTICAL PROCEDURE

A 50-70-mg sample was dissolved in 5 ml of dilute (1:2) H_2SO_4, 0.2 g Zn dust was added, and the mixture heated at 100° for 10 minutes. Reduction of the nitro group and evaporation of a large part of the water took place. If the substance did not dissolve in H_2SO_4, the mixture was stirred carefully, so as not to lose material on the walls of the flask. The flask was then cooled, another 5 ml of concentrated H_2SO_4 and 1.5-2 ml of chromic acid were added, and combustion was carried out by the previously described method. [4] The carbon content was determined from the weight of CO_2 absorbed in the Ascarite tube. To determine nitrogen, the sulfate solution, after combustion, was transferred to a flask with a safety trap, 50 ml 40% KOH solution was added, and NH_3 was distilled, as usual, into 0.1 N acid, the excess acid being titrated with 0.1 N alkali in the presence of 2-3 drops of methyl red.

TABLE

Simultaneous Determination of Carbon and Nitrogen in Compounds Containing Nitrogen in the Oxidized Form.

Compound	% C			% N		
	Found	Calc.	Δ	Found	Calc.	Δ
3-Carbethoxy-5-methoxy-2, 6-dimethyl-4-nitrobenzofuran, $C_{14}H_{15}NO_6$	57.15 57.44	57.33	-0.18 +0.11	4.75 4.58	4.78	-0.03 -0.20
3-Carbethoxy-5-methoxy-2-(*p*-nitrophenyl) benzofuran, $C_{18}H_{15}NO_6$	63.31 63.39	63.34	-0.03 +0.05	3.90 3.93	4.10	-0.20 -0.17
3, 5-Dinitrobenzoic acid, $C_7H_4N_2O_6$	39.51 39.71	39.63	-0.12 +0.08	13.22 13.09	13.21	-0.01 -0.12
m-Nitrophenol, $C_6H_5NO_3$	51.62 51.57	51.80	-0.18 -0.23	9.89 9.93	10.07	-0.18 -0.14
m-Nitroaniline, $C_6H_6N_2O_2$	52.02 51.95	52.17	-0.15 -0.22	20.18 20.31	20.28	-0.10 +0.03
4-Aminoazobenzene, $C_{12}H_{11}N_3$	72.82 73.11	73.07	-0.25 +0.04	21.62 21.57	21.31	+0.31 +0.26

The table shows, as examples, the results of carbon and nitrogen determinations in a single sample for only a few of the compounds analyzed.

It is evident from the data given that the "wet combustion" method with preliminary reduction can be used for the determination of carbon and nitrogen in nitro compounds. We must note, however, that this method cannot be used if CO or CS_2 are liberated during the reduction, or if acetic acid and other volatile organic compounds are formed.

It was to be expected that azo compounds would be reduced to amines under these conditions, and that it would be possible to determine carbon and nitrogen by subsequent combustion. The analysis of *p*-aminobenzene in fact showed results close to the calculated values (see Table). Hydrazones, which, in acid medium, hydrolyze to form hydrazine salts, can probably not be analyzed by this method.

When reduced and oxidized forms of nitrogen are present simultaneously, the proposed method can be used for their separate determination. By combustion with preliminary reduction, the total nitrogen content is obtained, e.g., in *p*-aminobenzene and *m*-nitroaniline (see Table). By another combustion without preliminary reduction, only the reduced form of nitrogen is determined [4], and the content of the oxidized form of nitrogen is obtained from the difference of the two determinations.

SUMMARY

A method has been developed for the simultaneous determination of carbon and nitrogen in nitro compounds, based upon the reduction of the nitro group by zinc dust in dilute sulfuric acid solution, with subsequent wet combustion by chromic and sulfuric acids by a previously described method. It was shown that in the simultaneous presence of the reduced and oxidized forms of nitrogen ($-NO_2$ or -N=N-), the method can be used for their separate determination.

LITERATURE CITED

[1] P. Fritsch, *Ann.* **294**, 79 (1897).
[2] H. Dieterle, *Arch. Pharm.* **262**, 35 (1924).
[3] C. N. Acharya, *J. Soc. Chem. Ind.* (*London*) **54**, 596 (1935); *Nature* **136**, 644 (1935).
[4] A. P. Terent'ev and B. M. Luskina, *Zhur. Anal. Khim.* **14**, 112 (1959).
[5] E. I. Aĭzenshtadt, *Zavodsk. Lab.* **9**, 233 (1940).
[6] A. Friedrich, E. Kühaas, and R. Schnürch, *Z. Physiol. Chem.* **216**, 68 (1933).
[7] P. R. W. Baker, *Analyst* **80**, 481 (1955).
[8] R. Belcher and M. K. Bhatty, *ibid.* **81**, 124 (1956).
[9] M. Krüger, *Chem. Ber.* **27I**, 609 (1894).
[10] T. S. Ma, R. E. Lang, and J. D. McKinley, Jr., *Mikrochim. Acta* **1957**, 368.
[11] V. I. Esafov, *Zavodsk. Lab.* **21**, 1160 (1955).
[12] W. E. Dickinson, *Anal. Chem.* **30**, 992 (1958).

10

*The Simultaneous Determination of C, Si and Al in Polymeric Organoaluminosiloxanes**

A. P. Terent'ev, B. M. Luskina,
and
S. V. Syavtsillo

Existing methods for the analysis of organosilicon compounds, based on combustion in a stream of oxygen, permit determination of C, H, and Si [1-2] or C, H, Si, and halogen (or S) [3] in a single sample. In the analysis of organosilicon compounds containing metal atoms, these methods are not suitable for the determination of both Si and metal. In such cases, the use of wet oxidation [4] is expedient.

Wet combustion, as a method for decomposing organosilicon compounds containing various metals and metalloids, can be used in very diverse cases, since, besides C and Si (after removing the latter), other elements remaining in the filtrate can be determined. The complete oxidation of various polymers can be performed by selecting suitable oxidizing mixtures and decomposition conditions.

We previously [5] suggested a method for oxidizing organosilicon compounds with sulfuric and chromic acids in a stream of oxygen to determine C and Si, and C, N, and Cu or Fe in organic compounds containing metals. In the present work it was

---

* *Zhur. Anal. Khim.* **16**, 635-638 (1961)

established that, for the decomposition of polymeric organoaluminosiloxanes by the wet combustion method, preliminary heating in concentrated H_2SO_4 is required. Apparently, this treatment ruptured the $>$Si-O-Al$<$ bond [6], and further oxidation of these siloxanes by chromic acid proceeded considerably more readily. The silicic acid formed was removed from the sulfate solution by filtration; gelatin solution was added to speed up filtration. The Si was converted further to sodium fluosilicate and was determined titrimetrically. [7] Al was determined in the filtrate after removing the Si. To separate Al from Cr, which is always found in significant quantity in the sulfate solution after combustion, we used electrolysis with a Hg cathode. [8] Cr was retained by the Hg, and Al remained in solution. Al was determined titrimetrically with Complexon III. [9] C was determined gravimetrically in the usual manner.

Since we did not have individual polymeric organoaluminosiloxanes, we established the accuracy of determination of Al in the following manner: Al was first determined in a technical polyphenylated aluminosiloxane by the method proposed (12.2% Al). Then a known volume of aluminum sulfate solution, prepared from AV 1 grade metallic Al, was introduced into reaction flask 3 (Fig. 1), evaporated almost to dryness, a sample of siloxane with known Al content was introduced, and oxidation and Al determination were then carried out as described below. The results obtained are given in Table 1.

The proposed method was used to determine C, Si, and Al in single samples of technical polymeric ethyl- and phenylaluminosiloxanes. (Table 2) The results obtained show the satisfactory agreement between parallel determinations.

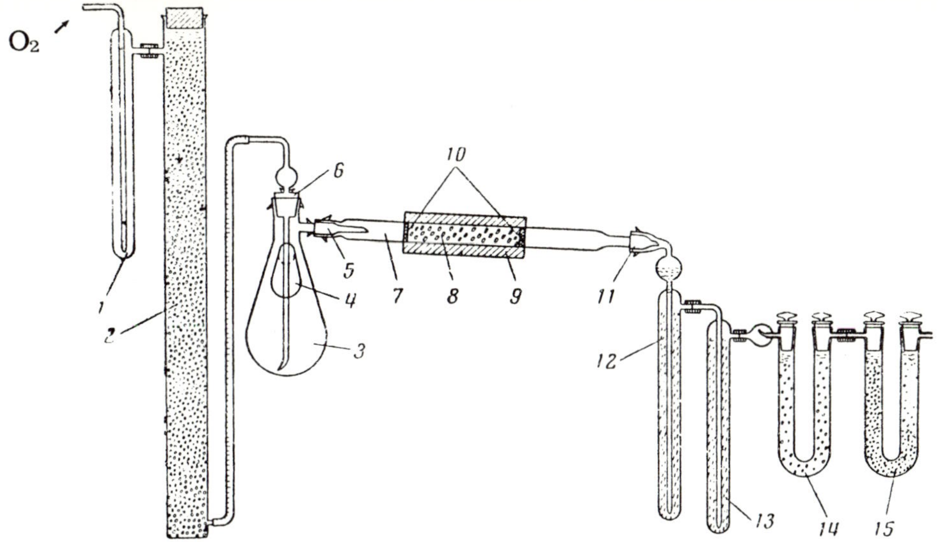

Fig. 1 Wet Combustion Apparatus. 1: Bubble counter with conc. H_2SO_4; 2: Ascarite column for O_2 purification; 3: reaction flask; 4: side bulb sealed into the flask for adding chromic acid; 5: ground joint; 6: ground stopper with O_2 inlet tube; 7: quartz tube (200 mm long, 8-10 mm diameter); 8: catalyst (Cr_2O_3 on pumice); 9: electric furnace at 700-750°; 10: loose asbestos stoppers; 11: ground joint; 12: absorber for halogens, containing pieces of glass rod moistened with an acetic acid solution of hydrazine (pH 6); 13: vessel containing pieces of glass rod moistened with conc. H_2SO_4; 14: Anhydrone tube; 15: tube with Ascarite and Anhydrone.

TABLE 1

Determination of Aluminum

Weight of polyphenyl-alumino-siloxane (mg)	Al (mg)				
	Contained in sample	Added	Total amount	Found	Δ
89.95	10.97	4.84	15.81	16.01	+0.20
90.64	11.06	2.42	13.48	13.59	+0.11
50.98	6.22	4.84	11.06	11.23	+0.17
50.53	6.14	2.42	8.56	8.52	-0.04
52.25	6.37	2.42	8.79	8.74	-0.05

TABLE 2

Simultaneous Determination of C, Si, and Al

Substance Analyzed	Weight (mg)	% C	% Si	% Al
Polyethyl-				
alumino-	49.10	15.3	15.2	14.6
siloxane	50.90	15.5	15.0	14.3
(Batch 1)				
Polyethyl-				
alumino-	50.88	16.7	12.2	17.2
siloxane	52.58	17.1	11.8	16.9
(Batch 9)				
Polyethyl-				
alumino-	48.30	15.1	14.0	13.9
siloxane	51.70	14.9	13.6	13.8
(Batch 21)				
Polyphenyl-				
alumino-	51. 20	35.7	12.0	15.5
siloxane	48.55	35.3	12.4	15.3

PROCEDURE

Into the dry reaction flask 3 (Fig. 1), containing the sample, is placed 5 ml conc. H_2SO_4*, and 1.5-2 ml chromic acid is placed in side bulb 4.** The flask is connected as shown in Fig. 1, a stream of oxygen is passed through for 2-3 minutes, and the weighed tube 15 is attached. While continuing the flow of oxygen, the flask is immersed for 10-15 minutes in a bath heated to 150-160°, and, after cooling, chromic acid is slowly added from side bulb 4. An evolution of bubbles of CO_2 is observed. The oxidation generally proceeds smoothly, without frothing. After the visible reaction is complete, the flask is again heated at the same temperature for 30 minutes.

The C content is determined from the weight of CO_2 absorbed in the Ascarite tube 15. Si and Al are determined in the contents of reaction flask 3 after combustion. The flask is disconnected, the silicic acid sticking to the walls is washed down with water and filtered off, and Si is determined titrimetrically, as described previously. [7]

The Al is determined as follows: After removing the Si, the Cr(VI) in the filtrate is reduced by heating with a few drops of methanol. The solution is transferred to a 200-ml measuring flask, an aliquot (50 ml)

---

*The H_2SO_4 is first heated for 40-50 minutes to decompose any possible organic impurities, and cooled in a flask with a $CaCl_2$ tube attached through a ground joint.

**70 g CrO_3 is dissolved in 100 ml water and boiled for 3-5 minutes. The cooled solution is filtered through a glass filter.

is taken and placed in an electrolyzer (Fig. 2) containing 30-40 ml metallic Hg and 50-60 ml water. The electrodes are connected to a d. c. source, and electrolysis is carried out at 5 amps and 22 volts for 100-120 minutes. The colorless solution obtained is siphoned from the electrolyzer into a conical flask, and the electrolyzer is washed 4-5 times with 5-7 ml water. The solution in the flask is evaporated down to about 50 ml, neutralized with 25% alkali solution (Congo paper), 1 ml conc. HCl and 20 ml 0.01 M Complexon III solution are added, and the mixture is

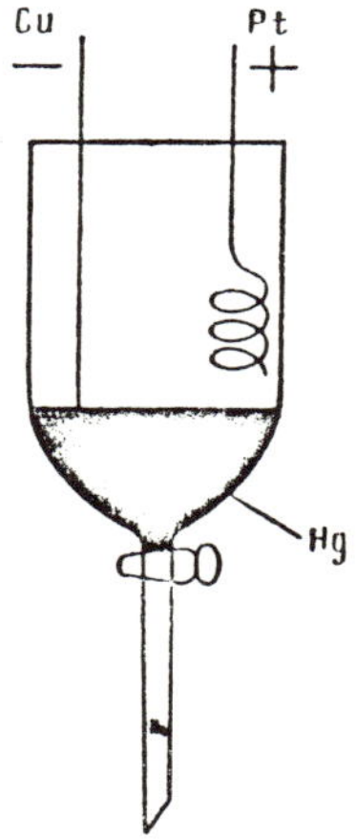

Fig. 2 Electrolyzer.

heated until boiling begins. On cooling, the solution is neutralized with 1 M ammonia solution, another 2-3 ml is added, and excess Complexon III is titrated rapidly with 0.01 M $ZnCl_2$ solution in the presence of

Chromogen Black ET-00.* A sharp change from blue to violet is observed at the end point. In the determination of Al, a control experiment should be carried out: 5 ml H_2SO_4 is added to 1.5-2 ml chromic acid, the Cr(IV) is reduced, and water is added to 200 ml. Fifty ml of the solution obtained is placed in the electrolyzer to remove Cr, and the solution is then taken through all the stages described above.

SUMMARY

A method has been developed for determining carbon, silicon, and aluminum in the same sample of polymeric organoaluminosiloxanes, by oxidizing the sample with a mixture of chromic and sulfuric acids at 150° in a stream of oxygen. Supplementary combustion of the possible products of incomplete combustion is carried out on chromic oxide, heated to 700-750°. The carbon content is determined gravimetrically, silicon is determined titrimetrically after conversion of the silicic acid to sodium fluosilicate, and aluminum is determined complexometrically.

LITERATURE CITED

[1] E. G. Rochow, *J. Am. Chem. Soc.* 70, 2170 (1948).

[2] V. A. Klimova, M. O. Korshun, and E. G. Bereznitskaya, *Zhur. Anal. Khim.* 11, 223 (1956).

---

*100 ml indicator is carefully mixed with 10 mg of finely ground NaCl [9], and the mixture is added on the tip of a microspatula.

[3] V. A. Klimova and E. G. Bereznitskaya, *ibid.* **11**, 292 (1956); **12**, 424 (1957).
[4] A. P. Kreshkov and V. A. Bork, *Usp. Khim.* **28**, 576 (1959).
[5] A. P. Terent'ev and B. M. Luskina, *Zhur. Anal. Khim.* **14**, 112 (1959).
[6] K. A. Andrianov and T. N. Ganina, *Izv. Akad. Nauk S. S. S. R., Otdel. Khim. Nauk* **1956**, 74.
[7] A. P. Terent'ev, S. V. Syavtsillo, and B. M. Luskina, *Zhur. Anal. Khim.* **16**, 83 (1961).
[8] W. F. Hillebrand, G. E. F. Lundell, et al.: "Laboratory Manual of Inorganic Analysis" [Russian Edition], Moscow: Goskhimizdat (1957), p. 153.
[9] G. Schwarzenbach: "Complexometry" [Russian Edition], Moscow: Goskhimizdat (1958), p. 115.

11

The Simultaneous Determination of C, Cl, N and Si; of C, P and Si; and C, Ti and Si in Organosilicon Compounds *

B. M. Luskina, A. P. Terent'ev,
and
S. V. Syavtsillo

Organosilicon compounds containing metals or nonmetals are characterized by great stability to oxidation and thermal action. The decomposition of such molecules, in order to determine their elemental composition, is difficult.

Dry combustion and wet oxidation methods are used for the analysis of organosilicon compounds. Methods based on the combustion of the sample in a stream of oxygen permit the determination of C, H, Cl, and Si. Wet oxidation methods, based upon the mineralization of the molecule of hetero-organic compounds with concentrated acids containing various oxidants or catalysts permit the determination of both silicon and a metal.

The proposed method of analysis [1] is based upon the oxidation of the sample with a mixture of concentrated sulfuric and chromic acids in an oxygen atmosphere at 150°C in an apparatus described previously.[3] The possible products of incomplete combustion are burned in a tube over Cr_2O_3, supported on pumice. C is determined gravimetrically as CO_2. Cl

* *Trudy Komissii Anal. Khim., Akad. Nauk SSSR, Inst. Geokhim. i Anal. Khim.* **13**, 3-7 (1963)

or Br is absorbed in a hydrazine hydrate solution, neutralized with acetic acid to pH 6, and is determined argentometrically. The direct determination of H is, of course, excluded, which is not decisive in a number of cases, but, on the other hand, many elements remaining in the reaction mixture can be determined simultaneously with C and halogen, including N, B, P, Si, metals, etc.

In the oxidation of organosilicon compounds, silicic acid is precipitated and is separated from the sulfate solution by filtration, which can be speeded up by adding gelatin solution. Si is then converted to sodium fluosilicate and determined volumetrically. [2]

From the examples in Table 1, it is seen that Cl was determined along with C and Si, and, in the case of dichloro(dimethylaminophenyl)methylsilane, four elements were determined simultaneously: C, Cl, N, and Si.

ANALYSIS OF POLY(ORGANOALUMINOSILOXANES)

The direct oxidation of polymeric organoaluminosiloxanes with chromic mixture does not proceed quantitatively under the conditions stated. Preliminary heating in concentrated sulfuric acid is necessary for their degradation by the wet oxidation method. In this treatment, the aluminosiloxane bond is ruptured, and further oxidation of the siloxanes formed proceeds readily. In this case, the combustion is carried out in the same apparatus. C is determined gravimetrically, Si volumetrically, and Al is determined in the filtrate after removal of Si and after separating the Al from Cr. [3]

The analysis of technical samples of polymeric ethyl- and phenylaluminosiloxanes showed satisfactory agreement between parallel determinations. (cf. [3], Table II)

ANALYSIS OF PHOSPHORUS-CONTAINING ORGANOSILICON COMPOUNDS

The choice of conditions for determining P by means of the wet oxidation method was made for individual organophosphorus compounds. The compound investigated was oxidized with a mixture of sulfuric and chromic acids in a stream of oxygen at 150°C in the apparatus described previously. [3] The P was oxidized to phosphoric acid. The direct determination of P in the presence of large amounts of Cr is difficult. As in the case of the determination of Al in poly(organoaluminosiloxanes), Cr was therefore removed by electrolysis, after which the solution contained only P, which was determined colorimetrically as the blue ammonium phosphomolybdate complex.

In the combustion of organophosphorous compounds containing Si by the above method, the C results, as a rule, were 1.5-2% low. In this case, more rigorous oxidation conditions were required for complete mineralization; viz., raising the temperature to 180°C. These conditions were favorable for the oxidation of both monomeric and polymeric compounds. Preliminary heating of the polymeric Si-P-organic compounds with sulfuric acid is as necessary as in the case of poly(organoaluminosiloxanes). [3]

C was determined gravimetrically, Si volumetrically, and P colorimetrically in the filtrate after removing the Si. Thus, C, Si, and P were determined simultaneously.

TABLE 1

Simultaneous determination of C, Cl, N, and Si.

Compound	% C		% Cl		% N		% Si	
	Calc.	Found	Calc.	Found	Calc.	Found	Calc.	Found
$(CH_3)_2N$ — C_6H_4 — $Cl_2(CH_3)Si$	46.16	46.19 45.72	30.28	30.31 29.90	5.98	6.26 6.13	11.99	12.42 12.47
$(C_2H_5)_2SiCl_2$	30.57	30.56 30.27	45.13	44.84 44.91	–	–	17.87	17.55 17.62
$ClC_6H_4SiCl_3$	29.29	29.05 29.20	57.65	57.38 57.46	–	–	11.42	11.13 11.23

ANALYSIS OF TITANIUM-CONTAINING ORGANOSILICON COMPOUNDS

Of these compounds, we have analyzed tetrakis–(triethylsiloxy)titanium (Table 2), which was burned with chromic mixture at 150°C. After determining C and separating Si, as described above, Ti was determined colorimetrically with H_2O_2, after first separating it from Cr by electrolysis with a mercury cathode.

TABLE 2

Determination of C, Si, and Ti in a Titanium-Containing Organosilicon Compound

Compound	mg Sample	% C Calc.	% C Found	% Si Calc.	% Si Found	% Ti Calc.	% Ti Found
Tetrakis (triethyl-siloxy)titanium	57.82	50.31	50.12	19.61	19.32	8.35	8.00
$C_{24}H_{60}O_4Si_4Ti$	44.85		50.24		19.28		8.10

Thus, wet oxidation, as a method of decomposition of organosilicon compounds containing various elements, can be used in the most varied cases. Other elements present in the molecule of the compound under analysis can also be determined along with carbon and silicon.

LITERATURE CITED

[1] A. P. Terent'ev and B. M. Luskina, *Zhur. Anal. Khim.* 14, 112 (1959).

[2] A. P. Terent'ev, S. V. Syavtsillo, and B. M. Luskina, *ibid.* **16**, 83 (1961).
[3] A. P. Terent'ev, B. M. Luskina, and S. V. Syavtsillo, *ibid.* **16**, 135 (1961).

12

The Determination of Nitrogen in Pyridine Compounds *

A. P. Terent'ev and B. M. Luskina

The nitrogen in heterocyclic compounds, and particularly pyridine derivatives, cannot always be determined quantitatively by the usual Kjeldahl method. The oxidation of these compounds frequently requires prolonged boiling with concentrated sulfuric acid, during which the ammonium sulfate formed sometimes decomposes, liberating elemental nitrogen. [1-4]

To speed up the decomposition process, one must add compounds that act as oxygen carriers or raise the b. p. of sulfuric acid. [5] Krüger [6] proposed adding potassium bichromate, which speeded up the determination considerably. Boiling in this case was carried out for 30 minutes. The method was tested by the authors on several derivatives of benzene, pyridine, and quinoline. However, according to our experiments, pyridine itself and ethylmethylpyridine do not give a quantitative yield of ammonia under these conditions.

The difficulty in oxidizing pyridine, in our opinion, is due to the fact that the pyridine ring is stable to the action of electrophilic reagents. Thus, the sulfonation of pyridine requires prolonged boiling with fuming sulfuric acid. The β-sulfopyridine formed is even

* *Zhur. Anal. Khim.* **17**, 227-230 (1962)

more stable to oxidation.[7, 8] To prevent sulfonation, one should evidently lower the heating temperature and decrease the concentration of sulfuric acid, after adding strong oxidants.

We first treated the sample with a concentrated aqueous solution of chromic anhydride in the presence of copper and potassium sulfates and mercuric oxide, and then added sulfuric acid. The reaction flask was equipped with a reflux condenser. The mixture was heated in a bath at 170-180° for 70-80 minutes. The decomposition time was reduced by further increasing the oxidizing properties of the mixture used, by replacing potassium sulfate by potassium persulfate. Thus, the mineralization of the difficultly oxidizable pyridine molecule was carried out with chromic acid and potassium persulfate in a dilute sulfuric acid medium, in the presence of copper sulfate and mercuric oxide. It was found that, under these conditions, pyridine and its derivatives are oxidized quantitatively in 30-35 minutes. The end of the reaction was established by the cessation of gas evolution. Then, as usual, excess potassium hydroxide was added, ammonia was distilled and determined by titration.

Table 1 shows some examples of these determinations.

It should be noted that the use of mercuric oxide as catalyst is inconvenient, since metallic mercury distils along with water vapor during the distillation of ammonia. It is therefore better to use selenium dioxide instead of mercuric oxide (Table 2).

There are various opinions concerning the determination of nitrogen in the presence of chlorine. Anderson and Schutte [9] explain it by the formation of ammonium chloride, which is oxidized to elemental nitrogen by potassium bichromate during the combustion. However, the experiments of Acharya [10] did

TABLE 1

Determination of Nitrogen in Pyridine Compounds in the Presence of Mercuric Oxide.

Compound	mg Sample	ml 0.1N HCl used	%N Found	%N Calc.	Δ
5-Ethyl-2-picoline, $C_8H_{11}N$	71.00	5.70	11.24	11.55	-0.31
	50.10	4.15	11.60		+0.05
Lutidine, C_7H_9N	55.10	5.11	12.99	13.07	-0.08
	56.40	5.26	13.06		-0.01
2, 6-Diacetylpyridine, $C_9H_9NO_2$	47.00	2.86	8.52	8.58	-0.06
	52.95	3.28	8.69		+0.11
Pyridoin, $C_{12}H_{10}N_2O_2$	47.40	4.34	12.82	13.08	-0.26

not confirm this assumption. Robertson [11] noted that, in the presence of chlorine, increasingly low results for nitrogen are obtained as the oxidation temperature increases. Upon prolonged boiling, no ammonia at all is obtained.

The reason for the incorrect results, in our opinion, lies in the oxidation conditions and in the formation of chromyl chloride. Low nitrogen values are obtained upon prolonged boiling with chromic mixture [9, 11], during which the ammonia formed evidently partly decomposes, liberating nitrogen. Chromyl chloride can probably also speed up the oxidation of ammonia to elemental nitrogen, if the combustion of the organic compound is carried out with concentrated sulfuric acid and dry potassium bichromate in the absence of water. [11] When dilute sulfuric acid or an

TABLE 2
Determination of Nitrogen in Pyridine and its Derivatives in the Presence of Selenium Dioxide.

Compound	mg Sample	ml 0.1 N HCl used	% N Found	% N Calc.	Δ
Pyridine, C_5H_5N	40.30	5.04	17.51	17.70	-0.19
	42.50	5.35	17.63		-0.07
2-Methyl-5-(3-methyl-3-acetoxy-1-butynyl) pyridine, $C_{13}H_{15}NO_2$	58.70	2.86	6.82	6.44	+0.38
	48.60	2.35	6.77		+0.33
2-Methyl-5-(3-hydroxy-3-methylbutyl)piperidine, $C_{11}H_{23}NO$	55.80	3.07	7.71	7.57	+0.14
	53.20	2.97	7.82		+0.25
1-[2-(2-methyl-5-pyridyl)ethyl]-2-methyl-3-propylindole, $C_{20}H_{24}N_2$	38.30	2.64	9.65	9.58	+0.07
	43.00	2.96	9.64		-0.06
6-Methyl-9-[2-(2-methyl-5-pyridyl) ethyl]-1, 2, 3, 4-tetrahydrocarbazole, $C_{21}H_{24}N_2$	30.70	2.04	9.30	9.20	+0.10
Pyridine-2, 6-dicarboxylic acid, $C_7H_5NO_4$	52.55	3.22	8.58	8.38	+0.20
	45.70	2.81	8.61		+0.23

aqueous solution of chromic acid is used [10, 12], chromyl chloride hydrolyzes, liberating hydrogen chloride, and correct nitrogen values are obtained.

The oxidation of chlorine-containing pyridine compounds by our method is carried out in a dilute sulfuric acid medium, which ensures complete hydrolysis of the chromyl chloride formed, and the presence of chlorine has no effect on the results of the nitrogen determination. Table 3 shows examples of the analysis of hydrochlorides of some pyridine derivatives.

It should be noted that the oxidation of a pyridine N-oxide is not accompanied by the formation of ammonia, and nitrogen cannot be determined by this method in such cases. In the analysis of 5-(2-ethylaminoethyl)-2-methylpyridine 1-oxide (Table 4), only the amine nitrogen is determined. In the case of 2-methyl-5-(4-pyrazolyl) pyridine, only the pyridine nitrogen is determined (Table 4), and the two other nitrogen atoms, in the five-member pyrazole ring, are liberated in the form of elemental nitrogen. Thus, the "reduced" form of nitrogen can be determined in the presence of the "oxidized" form by the method described (Table 4).

ANALYTICAL PROCEDURE

The oxidation is carried out in a round-bottomed flask with reflux condenser. A 50-70 mg sample is introduced into the flask, containing 2-3 ml of chromic acid solution (70 g CrO_3 in 100 ml H_2O) and a mixture of 0.25 g copper sulfate, 0.25 g selenium dioxide, and 5 g potassium persulfate. 10 ml sulfuric acid is then poured in through the reflux condenser. After the visible reaction is complete, the mixture is heated in a silicone bath at 170-180° until evolution of

TABLE 3
Determination of Nitrogen in Hydrochlorides of Pyridine Derivatives.

Compound		mg Sample	ml 0.1 N HCl used	%N Found	%N Calc.	Δ
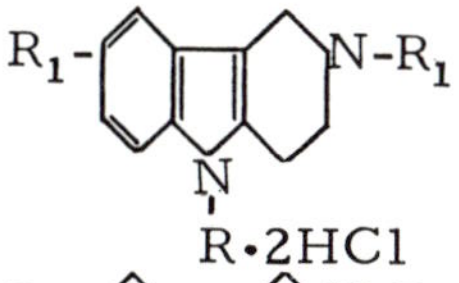	$C_{21}H_{27}Cl_2N_3$	41.90	3.30	11.03	10.71	+0.32
		41.40	3.24	10.92		+0.25
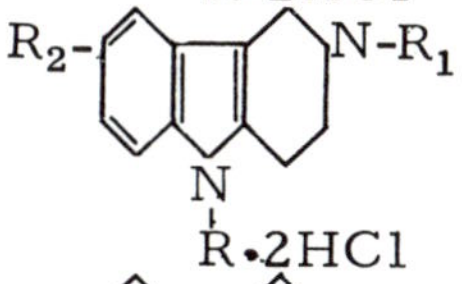	$C_{21}H_{27}Cl_2N_3O$	35.45	2.69	10.60	10.29	+0.31
		36.70	2.74	10.45		+0.12
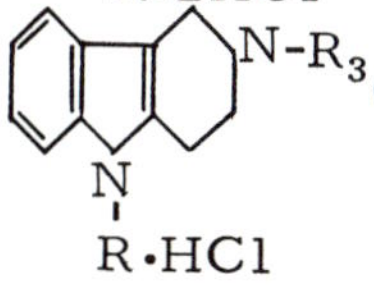	$C_{24}H_{32}ClN_3$	38.40	2.94	10.72	10.58	-0.14
		38.40	2.84	10.36		-0.20

R = $-CH_2-CH_2-$ 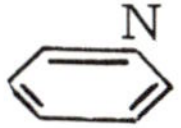 $-CH_3$; R_1 = $-CH_3$; R_2 = $-OCH_3$; R_3 = $-C_4H_9$

TABLE 4

Determination of the "Reduced" Form of Nitrogen in the Presence of the "Oxidized" Form.

Compound	mg Sample	ml 0.1 N HCl used	%N Found	%N Calc.	%N Δ
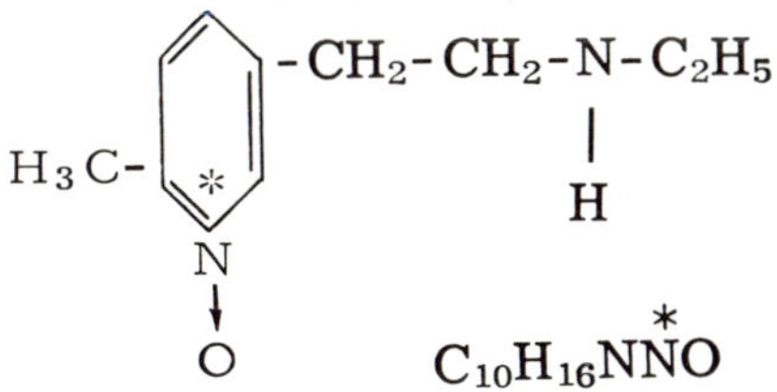 $C_{10}H_{16}N\overset{*}{N}O$	37.30	2.05	7.69	7.77	-0.08
	33.40	1.89	7.92		+0.15
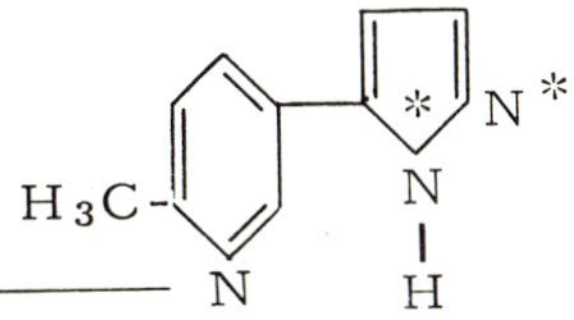 $C_9H_9N\overset{*}{N}_2$	42.70	2.61	6.78		-0.24
	42.40	2.66	8.56	8.80	-0.02

Note: N* is not converted to NH_3 and is not determined.

carbon dioxide is complete. The decomposition is usually complete within 30-35 minutes. With this oxidation method, there is no increase in the concentration of sulfuric acid. The reaction mixture is then transferred to a flask with a distillation adapter, excess alkali is added, and ammonia is distilled into a standard solution of acid, the excess of which is back-titrated with 0.05 N alkali solution in the presence of methyl red.

SUMMARY

A method for the determination of nitrogen in pyridine compounds is proposed, based on their oxidation with a mixture of potassium persulfate and chromic acid in sulfuric acid medium in the presence of copper sulfate and selenium dioxide at 170-180°. The determination takes one hour; the accuracy is ± 0.3% (absolute). Chlorine does not interfere with the determination.

LITERATURE CITED

[1] P. Fleury and H. Levaltier, *Bull. Soc. Chim. France* [4] **37**, 330 (1925).

[2] B. W. Grunbaum, et al., *Anal. Chem.* **27**, 384 (1955).

[3] P. R. W. Baker, *Analyst* **78**, 500 (1953).

[4] J. González Carreró, et al., *Inform. Quim. Anal.* (*Madrid*) **10**, 199 (1956).

[5] H. J. L. Meyer: "Analysis and Determination of the Structure of Organic Compounds" [Russian Edition], Khar'kov (1935), p. 131.

[6] M. Krüger, *Chem. Ber.* **27**, 609 (1894).

[7] S. M. McElvain and M. A. Goese, *J. Am. Chem. Soc.* **65**, 2233 (1943).
[8] B. A. Pavlov and A. P. Terent'ev: "Course in Organic Chemistry", Moscow (1958), p. 543.
[9] A. K. Anderson and H. S. Schutte, *J. Biol. Chem.* **61**, 57 (1924).
[10] C. N. Acharya, *Biochem. J.* **30**, 241 (1936).
[11] P. W. Robertson, *J. Chem. Soc.* **109**, 215 (1916).
[12] A. P. Terent'ev and B. M. Luskina, *Zhur. Anal. Khim.* **14**, 112 (1959).

13

*The Simultaneous Determination of C and N in Organic Compounds by "Wet Oxidation"**

A. P. Terent'ev and B. M. Luskina

Wet oxidation as a method of decomposing nitrogen-containing organic compounds for the simultaneous determination of carbon and nitrogen was used as early as 1897. [1]

We have shown [2, 3] that, in the oxidation of amines, nitriles, indoles, and pyrrolidones with a mixture of chromic and sulfuric acids, carbon is oxidized to CO_2 and determined gravimetrically, while nitrogen is converted to ammonium sulfate and is then determined by distilling off ammonia into a standard acid solution. The presence of nitro and azo groups in the compounds does not interfere with the determination of carbon and of reduced forms of nitrogen. The determination of azo and nitro nitrogen, however, requires preliminary reduction of the sample with zinc dust in a dilute sulfuric acid solution. The combustion is then carried out as usual, in the apparatus described previously. [4]

In this way, the number of nitro or azo groups can be determined together with carbon. Thus, for example, in nitroaniline, only the nitrogen in the amino group is determined by combustion without preliminary reduction, and the total nitrogen is determined

---

* *Trudy Komissii Anal. Khim., Akad. Nauk SSSR, Inst. Geokhim. i Anal. Khim.* **13**, 20-23 (1963)

after reduction. The number of nitro groups can be found from the difference between the two determinations.

Examples of the simultaneous determination of carbon and different forms of nitrogen are given in Table 1.

Pyridine compounds are incompletely oxidized by the method indicated above. As is well known, it is not always possible to determine nitrogen quantitatively in pyridines by the Kjeldahl method either, where the oxidation is carried out by boiling with concentrated sulfuric acid. The difficulty in this case is evidently that the pyridine ring is stable towards the action of electrophilic reagents.

The method of oxidation with chromic mixture has a number of significant advantages over the Kjeldahl method: the analysis is completed in 50-60 minutes, the process is conducted at the bench without a hood, and there is no danger of absorbing ammonia from the air in the laboratory; moreover, carbon can also be determined if necessary. To prevent sulfonation in the first stage, the oxidation was carried out in a medium of dilute sulfuric acid at a considerably lower temperature. [5]

PROCEDURE

Analysis is carried out in a round-bottomed flask with a reflux condenser. The sample (50-70 mg) is placed in the flask containing 2-3 ml of chromic acid (70 g CrO_3 dissolved in 100 ml water) and a mixture consisting of 0.25 g copper sulfate, 0.25 g selenium dioxide, and 5 g potassium persulfate. Ten milliliters of sulfuric acid is then poured in through the condenser. Upon completion of the reaction, the

TABLE 1

Determination of Carbon and Different Forms of Nitrogen

Compound Analyzed	% C Found	% C Calc.	% C Δ	%N Found	%N Calc.	%N Δ
p-Aminobenzoic acid ($C_7H_7NO_2$)	61.56 61.42	61.31	+0.25 +0.11	10.13 10.10	10.21	-0.08 -0.11
2-Oxo-1, 1-cyclohexanedipropionitrile ($C_{12}H_{16}N_2O$)	70.50 70.42	70.56	-0.06 -0.14	13.61 13.56	13.72	-0.11 -0.16
Ethyl 5-acetoxy-2-methyl-1-phenyl-3-indolecarboxylate ($C_{20}H_{19}NO_4$)	71.41 71.26	71.20	+0.21 +0.06	4.02 4.10	4.15	-0.13 -0.05
2-Methyl-1-[2-(phenoxy-1-phenoxymethylethoxy)ethyl]pyrrolidine ($C_{22}H_{29}NO_3$)	74.05 74.26	74.33	-0.28 -0.07	3.66 3.94	3.94	-0.28 0.00
2, 4-Dinitrobenzoic acid ($C_7H_4N_2O_6$)	39.51 39.71	39.63	-0.12 +0.08	13.22 13.09	13.21	+0.01 -0.12
m-Nitroaniline ($C_6H_6N_2O_2$)	52.22 52.19	52.17	+0.05 +0.02	10.18 10.08	10.14*	+0.04 -0.06
m-Nitroaniline ($C_6H_6N_2O_2$)	52.02 51.95	52.17	-0.15 -0.22	20.18** 20.31**	20.28	-0.10 +0.03
p-Phenylazoaniline ($C_{12}H_{11}N_3$)	73.05 72.90	73.07	-0.02 -0.17	7.00 7.23	7.10*	-0.10 +0.13
p-Phenylazoaniline ($C_{12}H_{11}N_3$)	72.82 73.11	73.07	-0.25 +0.04	21.62** 21.57**	21.31	-0.31 +0.26

* This calculated value is the amino nitrogen content.

** The total nitrogen content was determined after reduction with zinc dust.

TABLE 2

Determination of Nitrogen in Pyridine Compounds.

	mg	%N		
Compound Analyzed	Sample	Found	Calc.	Δ
α, α, 6-Trimethyl-2-piperidinepropanol ($C_{11}H_{23}NO$)	55.80	7.71	7.57	+0.14
	53.20	7.82		+0.25
5-Ethynyl-2-picoline (C_8H_7N)	76.30	11.95	11.97	-0.02
	50.90	11.94		-0.03
2, 5-Dimethyl-1-[2-(6-methyl-2-pyridyl) ethyl]indole ($C_{18}H_{20}N_2$)	46.70	10.22	10.60	-0.38
	36.40	10.16		-0.44
2, 6-Pyridinedicarboxylic acid ($C_7H_5NO_4$)	52.55	8.58	8.38	+0.20
	45.70	8.61		+0.23
Pyridine (C_5H_5N)	40.30	17.51	17.70	-0.19
	42.50	17.63		-0.07

mixture is heated on a silicone bath at 170-180°C until the evolution of CO_2 ceases. The decomposition is usually completed in 30-40 minutes. In this method of oxidation, no increase in the concentration of sulfuric acid takes place. The reaction mixture is then transferred to a flask with an adapter, excess alkali is added, and ammonia is distilled off into a standard acid solution, whose excess is back-titrated with 0.05 N alkali. Some examples of the analysis of pyridine compounds are given in Table 2.

LITERATURE CITED

[1] P. Fritsch, *Ann.* **294**, 79 (1897).
[2] A. P. Terent'ev and B. M. Luskina, *Zhur. Anal. Khim.* **14**, 112 (1959).
[3] B. M. Luskina, Dissertation, Moscow State University (1959).
[4] A. P. Terent'ev, B. M. Luskina, and S. V. Syavtsillo, *Zhur. Anal. Khim.* **16**, 635 (1961).
[5] A. P. Terent'ev and B. M. Luskina: "Summary of Reports of the Conference on Organic Analysis", Moscow: Publishing House of Moscow State University (1961), p. 16.

14

*The Simultaneous Determination of P and Si in Organic Compounds**

B. M. Luskina, A. P. Terent'ev
and
S. V. Syavtsillo

The literature contains only very limited information on the analysis of organic compounds containing silicon and phosphorus. Fennell and Webb [1] determined Si and P in a single sample for some high-boiling compounds containing both elements. The authors used two methods of "mineralization". In one of these, they used wet oxidation with a mixture of sulfuric and nitric acids. The precipitated silicic acid was filtered off, dried, ignited, and weighed. Phosphorus was determined gravimetrically in the solution after removal of the Si. In the second method, the compound to be analyzed was fused with sodium peroxide in a bomb. The subsequent determination of Si and P was also carried out gravimetrically. It should be noted that the determination of P and Si is lengthy in both cases.

In our earlier method [2] for the simultaneous determination of C, Si, and P in organic compounds containing Si and P, "mineralization" was carried out with a mixture of chromic and sulfuric acids in a stream of oxygen at 180°. The Si was precipitated as silicic acid, which was filtered off and determined

* *Zhur. Anal. Khim.* **17**, 639-641 (1962)

titrimetrically. [3] The P was determined colometrically in an aliquot part of the solution after electrolytic determination of the chromium.

In the present work, we used $K_2S_2O_8$ or H_2O_2 in concentrated H_2SO_4 to mineralize the organic compounds containing Si and P, which permitted a simplification and speed-up of the analysis.

EXPERIMENTAL

The "mineralization" of organic compounds containing Si and P was carried out as follows: A weighed sample of the substance was dissolved in concentrated H_2SO_4 with gentle warming, $K_2S_2O_8$ was added, and the mixture was heated on an electric hot plate until a homogeneous solution was obtained. The precipitated silicic acid was filtered off, dissolved in alkali, and determined by titration. [3] We assumed that P could be determined in the filtrate after removal of Si, in the form of the blue complex of ammonium phosphomolybdate. However, we were unable to obtain the "phosphomolybdenum" blue under these conditions, which is evidently explained by the presence of residual traces of undecomposed oxidant. It was possible to remove this difficulty by precipitating P as ferric phosphate. This was done by adding ferric ammonium sulfate solution to the filtrate after removal of Si, and then ammonia to complete the precipitation of iron. The P was coprecipitated with the ferric hydroxide. It should be noted that iron does not interfere with the further determination of P.

Under these conditions we analyzed individual organic silicon-phosphorus compounds containing considerable quantities of P. In this case, the P determination was carried out on an aliquot of the solution obtained after removal of Si. The results are shown in Table 1.

TABLE 1

Compound*	%Si Found	%Si Calc.	%Si Diff.	%P Found	%P Calc.	%P Diff.
Diethyl (diethoxymethylsilyl)methyl phosphate, $C_{10}H_{25}O_6PSi$	9.00 9.30	9.35	-0.35 -0.05	10.15 10.25	10.29	-0.14 -0.04
Bis(trimethylsilylmethyl) methylphosphonate, $C_9H_{25}O_3PSi_2$	20.87 21.20	20.93	-0.06 +0.27	12.06 11.81	11.81	+0.25 0.0
(Ethoxydimethylsilyl)-methyl dimethylphosphinate, $C_7H_{19}O_3PSi$	12.96	13.35	-0.39	14.66	14.73	-0.07

* The compounds were kindly provided by K. A. Andrianov.

We also oxidized the organic silicon-phosphorus compounds with H_2O_2. 1-2 ml of 70% H_2O_2 solution or 3-4 ml of phosphorus-free perhydrol is required to decompose 30-90 mg of sample. The H_2O_2 should be added gradually, in small portions, so that the decomposition does not proceed too violently. Comparative P determinations carried out by decomposition with $K_2S_2O_8$ and H_2O_2 showed that both methods give corresponding results.

ANALYTICAL PROCEDURE

A 30-90 mg sample is placed in a 100-ml Erlenmeyer flask containing 5 ml of concentrated H_2SO_4, the mixture is heated slightly, 2 g of $K_2S_2O_8$ is added to the cooled solution, and it is then heated for 10-15 minutes on a hot plate. A chemical funnel is inserted in the flask to prevent spattering. The flask is then again cooled to 75-50°, another 2-3 g of $K_2S_2O_8$ is added, and it is again heated for 10-15 minutes. The decomposition of the sample takes 30-40 minutes. At the end of the "mineralization," the solution becomes colorless, with a suspension of silicic acid.

When using perhydrol, the decomposition is also carried out in two stages, adding 2 ml each time. The total heating time in this case is 20-25 minutes.

After decomposing the sample by one of the above methods, the solution is cooled, 20-30 ml of water is added, the mixture is boiled at 110-120° for 5 minutes, then cooled to 70-60°, 1-1.5 ml of 2% gelatin solution added, and left for 3-5 minutes to coagulate the precipitate. The precipitated silicic acid is filtered off through a loose filter, placed in a polyethylene funnel, and washed 3-4 times with water.

The precipitate on the filter is dissolved in alkali. The solution obtained is neutralized, saturated with KCl, and a neutral solution of NH_4F and a known volume of 0.1 N HCl are added; the excess of the latter is titrated with alkali solution. [3]

Phosphorus is determined in the filtrate after removal of Si. When $K_2S_2O_8$ is used for the oxidation, the procedure is as follows: 1 ml of 3.5% ferric ammonium sulfate solution is added to the solution, followed by NH_4OH solution to complete the precipitation of iron, with which P coprecipitates. The precipitate is filtered off, washed with cold water, and the filtrate is discarded. The precipitate on the filter is dissolved in a minimum amount (2-3 ml) of 8 N HCl, the filter is carefully washed 3-4 times with distilled water, the solution is collected in the flask in which the iron was precipitated and is then transferred to a 200-ml measuring flask and made up to the mark with water.

In the case of "mineralization" with H_2O_2, the P determination is simplified. 0.1 N $KMnO_4$ is added dropwise to the solution after removal of Si, to decompose the H_2O_2. Then the solution is again transferred to a measuring flask.

An aliquot of the solution obtained by one of the above methods is taken from the measuring flask, made up to 35-40 ml with water, 2.5 ml of ferric ammonium sulfate solution is added, the mixture is neutralized with NH_4OH, 2 ml HCl is added, and the iron is reduced to the divalent state with 6 ml of 20% Na_2SO_3 solution with heating. The solution is cooled, another 6 ml HCl (1:3) and 4 ml of 5% ammonium molybdate solution are added, and the optical density of the blue solution of ammonium phosphomolybdate is measured. [4]

SUMMARY

A method has been developed for the simultaneous determination of silicon and phosphorus in organic compounds, based upon mineralization of the sample with potassium persulfate or perhydrol in concentrated sulfuric acid on heating. Silicon is precipitated and is then determined titrimetrically. The phosphorus in the filtrate after removal of the silicon is determined photometrically in the form of the blue complex of ammonium phosphomolybdate. The determination takes two hours.

LITERATURE CITED

[1] T. R. F. W. Fennell and J. R. Webb, *Talanta* **2**, 389 (1959).

[2] B. M. Luskina, A. P. Terent'ev, and S. V. Syavtsillo, *Trudy Komissii Anal. Khim.*, *Akad. Nauk S.S.S.R.*, *Inst. Geokhim. i Anal. Khim.* **13**, 3 (1963).

[3] A. P. Terent'ev, S. V. Syavtsillo, and B. M. Luskina, *Zhur. Anal. Khim.* **16**, 83 (1961).

[4] A. M. Dymov: "Technical Analysis of Ores and Metals", Moscow (1949), p. 276.

15

The Determination of Carbon in the Organic Material Present in Industrial Effluent that Contains Organosilicon Compounds *

A. P. Terent'ev, B. M. Luskina,
and
S. V. Syavtsillo

Effluent from the manufacture of organosilicon products may contain a number of organic substances, such as alcohols, toluene, and various organosilicon compounds. Constant control of the level of these impurities requires simple and available methods. We have already published methods for the determination of alcohols and organosilicon compounds. [1] The present paper describes a method for determining the total carbon content of the organic compounds.

The large number of methods for determining carbon in industrial effluent can be divided into three groups. The first comprises those methods in which carbon is determined from the oxidation equivalent. In this case, oxidation is carried out with a mixture of sulfuric and chromic acids, and the unconsumed amount of the latter is measured. [2] Kleinert [3], in his critical remarks on the wet combustion method, showed that determination from the excess of oxidant is permissible only in cases where the chemical individuals or mixtures contain hydrogen and oxygen in the ratio 2:1. In all other cases, when hydrogen and

* *Zhur. Anal. Khim.* 18, 639-643 (1963)

oxygen are not present in this ratio, oxidizability cannot be used as a measure of the carbon content. These methods are useful when it is not necessary to know the exact carbon content. Kleinert's remarks are correct. The determination of carbon from the unconsumed oxidant cannot give accurate results, since the oxidation does not always go to completion, and the oxidant is often decomposed to other substances. Moreover, it is known that, with increase in the concentration of chromic acid and in the heating temperature, decomposition of the acid increases and, upon prolonged heating with sulfuric acid, may even become complete. Consequently, this method requires the introduction of corrections for the decomposition of the oxidant. To avoid large corrections, oxidation is frequently carried out at relatively low temperatures (100-120°C), in the presence of a minimum amount of oxidant; however, only a small number of organic compounds are completely oxidized by this process.

A second group [4-8] comprises those methods in which the carbon is determined from the CO_2 formed upon oxidation with a mixture of chromic and sulfuric acids. In this case, as in the determination of the oxidation equivalent, incomplete oxidation is possible as a result of the formation of CO, acetic acid, etc..

The third group comprises methods of wet oxidation with dry combustion [9-11]. Such combination methods permit the reliable combustion of even difficultly combustible organosilicon compounds.

To determine the total carbon content in the effluent from the production of organosilicon compounds, we have used a combination method of combustion developed by us previously [11], based on the oxidation of organic substances by a mixture of sulfuric and chromic acids in a stream of oxygen under heating, with additional combustion of the gaseous reaction products over ignited chromium sesquioxide. The CO_2 formed was absorbed by Ascarite and weighed. Using 1-5 ml of effluent for analysis, the sensitivity of the determination was 1 g/l, which is inadequate. The sensitivity of the carbon determination can be increased by replacing the gravimetric finish by a titrimetric one.

An important point in the titrimetric determination of CO_2 is the establishment of the conditions for the complete absorption of CO_2. CO_2 is poorly soluble in water, with an absorption coefficient of 0.878 at 20°C [12]; consequently, a large excess of solution is required for complete absorption. CO_2 dissolves better in organic solvents, such as bipiperidines [13], ethanolamine [14], ethylenediamine [15], aniline [16], pyridine [17], etc.

A fairly large number of apparatuses to ensure a sufficiently complete absorption of CO_2 have been proposed. [7, 18-20] However, the use of freshly distilled solvents and special devices for the absorption of CO_2 introduces many complications into the technical control of effluent purification.

The main reason for the incomplete absorption of CO_2 by the absorbent liquid, particularly in a gas stream, is the formation of mists; i.e., mixtures of water vapor and gas. To remove such mists, Fedoseev and Sobko [21] have used a small glass tube with a finely crystalline halide powder. One of the authors had previously carried out the absorption of sulfur oxides in an evacuated vessel containing the absorbent solution. [22]

For the complete absorption of CO_2, we have used an evacuated flask containing an aqueous solution of $Ba(OH)_2$, into which oxygen and the CO_2 formed are

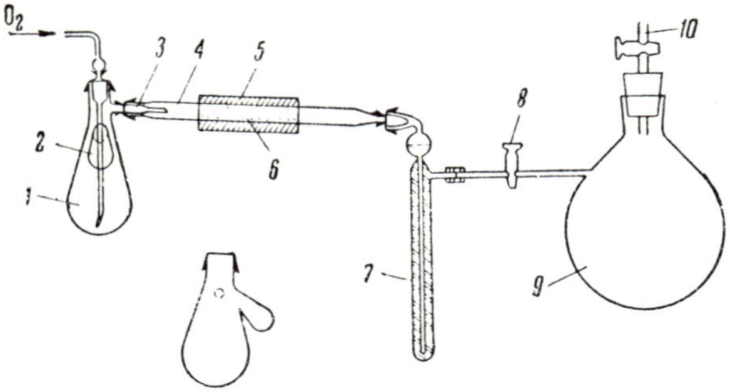

Fig. 1 Apparatus for the titrimetric determination of carbon in industrial waters by the "wet combustion" method. 1: Reaction flask (50-60 ml capacity); 2: side bulb; 3: ground joint; 4: quartz tube (length 200 mm, diameter 10 mm); 5: electric furnace at 700-750° C; 6: catalyst: Cr_2O_3 on pumice; 7: absorption tube containing pieces of glass rod moistened with an acetic acid solution of hydrazine hydrate, to absorb halogens; 8: stopcock; 9: 500-ml flask; 10: stopcock.

drawn. With this method of gas absorption, no gas escape is observed, and there is no need to control the oxygen flow rate. The gaseous reaction products are completely retained in the flask, the water vapor gradually condenses, and the CO_2 is freed from the water vapor and quantitatively absorbed.

It should be taken into account that the effluent may contain dissolved CO_2 or carbonates. Therefore, before the combustion, we acidified the water sample with sulfuric acid and passed a stream of oxygen through to displace CO_2.

EXPERIMENTAL PART

Carbon was determined in the apparatus shown in Fig. 1. The reliability of the method was tested with aqueous solutions of organic compounds, prepared by dissolving the chemical individuals in known volumes of distilled water. To determine carbon, aliquot parts of the solution were taken. The water-insoluble toluene and octaethyltetracyclosiloxane were analyzed as separate weighed samples placed in distilled water. The results are shown in Table 1, from which it is

TABLE 1

Determination of Carbon in Aqueous Solutions of Organic Compounds.

Compound dissolved in the water	ml water taken	mg C Taken	mg C Found	Difference (mg)	% Error (relative)
Methanol	1	1.80	1.65	-0.15	8.3
	5	9.00	8.92	-0.08	0.9
Isopropanol	1	6.70	6.03	-0.67	10.0
	1	6.70	6.53	-0.17	2.5
Toluene	2	4.35	4.10	-0.25	5.7
	2	3.90	3.65	-0.25	6.4
Isobutanol	1	1.28	1.16	-0.12	9.3
	2	2.56	2.80	+0.24	9.4
Octaethyltetra-	2	3.30	3.05	-0.25	7.5
cyclosiloxane	3	2.93	3.10	+0.17	5.7

seen that the accuracy is no greater than 10% (relative). The sensitivity of the determination is 100-150 mg/l of carbon. The effluent after purification generally contains little organic carbon, and the titration in this case was carried out with 0.01 N solutions. This increased the sensitivity to 10-15 mg/l. Results for some samples of waste waters are given in Table 2, which shows satisfactory agreement between parallel determinations.

TABLE 2

Determination of Carbon in Waste Water.

Sample No.	C found (mg/l)	Deviation from mean (mg/l)	Sample No.	C found (mg/l)	Deviation from mean (mg/l)
1	367.6	23.0	3	1103.0	30.0
	321.7			1163.0	
2	581.3	15.6	4 *	36.0	8.0
	612.5			28.0	

*The CO_2 was absorbed by means of 0.01 N $Ba(OH)_2$ solution.

PROCEDURE

10 ml of sulfuric acid* is poured into the round-bottomed reaction flask 1 (see Fig. 1), and 1-5 ml of the water to be analyzed is added with a pipette. 2-3

* The sulfuric acid (sp. gr. 1.84) is first heated for 40-50 minutes to remove organic impurities and cooled in a flask with a calcium chloride tube attached through a ground glass joint.

ml of chromic acid solution* is introduced into the side bulb 2, by means of a funnel with a bent tube. Flask 1 is connected as shown in Fig. 1, and a stream of oxygen is passed through for 5 minutes to displace air and dissolved CO_2 from the analytical sample of water. 20 ml of 0.1 N $Ba(OH)_2$ solution (or 0.01 N solution, if the water is analyzed after purification) is added, the flask is evacuated to a residual pressure of 50-100 mm, and connected to the apparatus as shown in Fig. 1. The oxygen flow is regulated by means of stopcock 8, and is maintained at such a rate that it is no longer possible to count the bubbles.

The combustion is carried out as follows: Chromic acid solution is added dropwise from the side bulb 2 to the flask 1, by rotating the bulb around the ground joint 3, which is lubricated with sulfuric or phosphoric acid. After the visible decomposition is complete, flask 1 is placed in a bath heated to 150-160°C, and held at this temperature under the oxygen stream for 20-25 minutes. The gaseous products of incomplete combustion are burned in tube 4 over ignited Cr_2O_3 [23], and then pass into the absorption tube 7, filled with pieces of glass rod moistened with an acetic acid solution of hydrazine hydrate neutralized to pH 6, which retains halogen. The CO_2 liberated is absorbed by $Ba(OH)_2$ solution in flask 9. After the combustion is complete, a rapid stream of oxygen is passed through until absorption ceases, stopcock 8 is closed, flask 9 is left for 10-15 minutes, and the excess $Ba(OH)_2$ solution is then titrated with an HCl solution of corresponding normality, in the presence of phenolphthalein. A control experiment is carried out under the same conditions.

---

* 70g CrO_3 is dissolved in 100 ml water, boiled for 3-5 minutes, and the cooled solution is filtered through a glass frit.

SUMMARY

A method for the determination of carbon in effluent has been developed, based upon the wet oxidation of organic compounds by chromic and sulfuric acids with an additional combustion over chromium sesquioxide, and a titrimetric determination of CO_2. Under these conditions, volatile impurities are also oxidized quantitatively.

To eliminate the mist formed in the combustion, we have used an evacuated round-bottomed flask, in which the CO_2 is absorbed and the titration is carried out.

A determination takes 40-50 minutes. The error is 10% (relative), with a sensitivity of 10-15 mg of carbon per liter of effluent.

LITERATURE CITED

[1] B. M. Luskina, S. V. Syavtsillo, et al., *Plasticheskie Massy* **1963**, No. 5, 61.

[2] W. A. Moore, R. C. Kroner, and C. C. Ruchhoft, *Anal. Chem.* **21**, 953 (1949).

[3] H. Kleinert, *Kühn-Arch.* **69**, 603 (1955).

[4] J. W. White and F. J. Holben, *Ind. Eng. Chem.* **17**, 83 (1925).

[5] H. Bach, *Z. Anal. Chem.* **91**, 259 (1933).

[6] E. Sauer and G. Rak, *ibid.* **114**, 182 (1938).

[7] A. A. Houghton, *Analyst* **70**, 118 (1945).

[8] Yu. Yu. Lur'e and A. I. Rybnikova: "Methods for the Chemical Analysis of Industrial Waste Waters", Moscow (1953), p. 134.

[9] W. P. Pickhardt, A. N. Oemler, and J. Mitchell, Jr., *Anal. Chem.* **27**, 1784 (1955).

[10] J. Rutschmann and W. Schöniger, *Helv. Chim. Acta.* **40**, 428 (1957).

[11] A. P. Terent'ev and B. M. Luskina, *Zhur. Anal. Khim.* **14**, 112 (1959); A. P. Kreshkov, V. T. Shemyatenkova, et al., *ibid.* **15**, 635 (1960).
[12] "Handbook of Chemistry and Physics", 44th Ed. (1962-1963), p. 1706.
[13] R. B. Evans and D. W. Parkes, *J. Soc. Chem. Ind.* **57**, 302 (1938).
[14] A. L. Shneerson and A. G. Leibush, *Zh. Prikl. Khim.* **22**, 553 (1949).
[15] R. W. Swick, D. L. Buchanan, and A. Nakao, *Anal. Chem.* **24**, 2000 (1952).
[16] C. J. Van Nieuwenburg and L. A. Hegge, *Anal. Chim. Acta* **5**, 68 (1951).
[17] L. Bloin and L. Edelhausen, *ibid.* **13**, 120 (1955).
[18] A. G. Filippova, *Zavodskaya Lab.* **8**, 497 (1939).
[19] M. I. Kulenok, *ibid.* **17**, 1322 (1951).
[20] B. F. Ormont and V. I. Smirnova, *ibid.* **18**, 263 (1952).
[21] P. N. Fedoseev and M. Ya. Sobko, *Zhur. Anal. Khim.* **10**, 323 (1955); **14**, 118 (1959).
[22] P. N. Fedoseev and R. M. Lagoshnaya, *Zhur. Anal. Khim.* **9**, 37 (1954).
[23] A. P. Terent'ev, B. M. Luskina, and S. V. Syavtsillo, *Zavodskaya Lab.* **25**, 288 (1959).

16

*The Simultaneous Determination of Si, P and Al in Organic Compounds**

B. M. Luskina, A. P. Terent'ev,
and
N. A. Gradskova

Analysis of organosilicon polymers containing metal and metalloid atoms along with silicon and oxygen in the molecular chain has not undergone any extensive development. These compounds are usually resistant to degradation and thermal decomposition, and quantitative conversion of the elements being determined to convenient forms for analysis is difficult.

The methods for the determination of P and Si or Al and Si in phosphoro- or alumino-silico-organic compounds previously developed by us [1, 2] cannot be combined and used for the analysis of compounds containing silicon, phosphorus, and aluminum. In the wet oxidation of organosilicon compounds containing another metal or metalloid, the elements being determined (Al, P, Ti, and others) remain in solution in the form of sulfates.[3] In the oxidation of compounds containing silicon, phosphorus, and aluminum, a precipitate of aluminum phosphate is formed along with the precipitate of silicic acid, but neither aluminum nor phosphorus, which are usually found in a 1:1 ratio in the molecules of these

**Zhur. Anal. Khim.* 19, 1251-1253 (1964)

compounds, is precipitated completely. Quantitative precipitation of one of these elements requires a tenfold excess of the other. [4]

The determination of aluminum and phosphorus by addition of the amounts found in the precipitate and determined in the filtrate after separation of the precipitate is lengthy and laborious. Moreover, aluminum phosphate may have different compositions, depending on the conditions of its preparation (temperature, acidity, etc.), e.g., $AlPO_4.2H_2O$, $Al(H_2PO_4)_3$, $Al_2(HPO_4)_3.8H_2O$, $Al_4(P_2O_6)_3.23H_2O$, etc. [5]

It is known that aluminum phosphates salt out in concentrated sulfuric acid solution, and go back into solution upon dilution with water. We have selected the optimum conditions for dissolving a precipitate of aluminum phosphate both for solutions of aluminum and phosphorus salts (Table 1) and for individual phosphoro-alumino-organic compounds (Table 2). It was found that in order to dissolve the aluminum phosphate formed, the sulfuric acid solution obtained after oxidation must be diluted with water in the ratio 1:3 by volume and boiled for 10-15 minutes.

TABLE 1

Determination of Aluminum and Phosphorus in Solutions of Their Salts.

mg Al			mg P		
Taken	Found	Δ	Taken	Found	Δ
4.08	4.15	+0.07	–	–	–
–	–	–	2.14	2.10	-0.04
10.0	9.65	-0.35	4.0	4.15	+0.15
4.08	4.10	+0.02	2.14	2.16	+0.02

The "mineralization" of polymeric organosilicon

compounds containing aluminum and phosphorus [6, 7] was carried out by means of potassium persulfate and heating with sulfuric acid. It has been established that mineralization of polymeric compounds is more rapid and more complete if they are dissolved in sulfuric acid. Organic compounds containing silicon, phosphorus, and aluminum behave differently toward sulfuric acid. Some of them dissolve readily even in the cold; others do not dissolve even upon heating. In the latter case, when the oxidant is added, the sample under analysis is covered with an oxide film, and decomposition ceases even after prolonged boiling of the solution. To improve the dispersion of the polymers in the sulfuric acid, they were first dissolved in a minimum amount of diethyl ether. The ether solution obtained was run into concentrated sulfuric acid and heated for 3-5 minutes to break down the aluminosiloxane bond. [1]

Upon subsequent addition of oxidant, the decomposition proceeded rapidly (20-30 minutes). Completeness of mineralization was judged by the external appearance of the solution (colorless, with a suspension of silicic acid gel). Under the conditions described, compound 1 (Table 2) was analyzed for Si, P, and Al. The results obtained are close to the theoretical values.

In the analysis of other polymethylphenylaluminosiloxanes containing phosphorus, for example with m less than 13, the analytical results obtained by the method described were low. Evidently, compounds of this type with a shorter siloxane chain have a higher thermal stability and are harder to degrade. Such phenylated polymethylaluminophosphorosiloxanes are completely oxidized under our conditions when heated for 2-2.5 hours with potassium persulfate. The oxidation time is considerably reduced if the sample weight is cut in half. Analytical results for these compounds are shown in Table 2.

TABLE 2

Determination of Silicon, Aluminum, and Phosphorus in Organic Compounds

Sample No.	Compound	% Si			% Al			% P		
		Found	Calc.	Δ	Found	Calc.	Δ	Found	Calc.	Δ
1	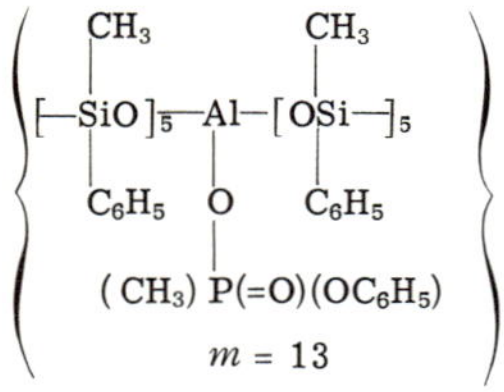	18.14	18.00	+0.14	2.14	1.72	+0.42	1.90	1.98	-0.08
		17.90		-0.10	2.17		+0.45	1.80		-0.18
2	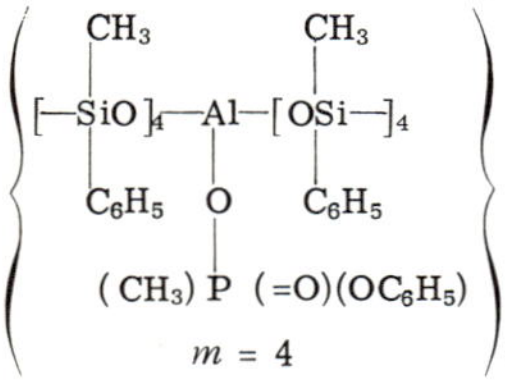	17.94	17.44	+0.50	2.00	2.09	-0.09	2.00	2.41	-0.41
		17.50		+0.06	2.21		+0.12	2.02		-0.39
3	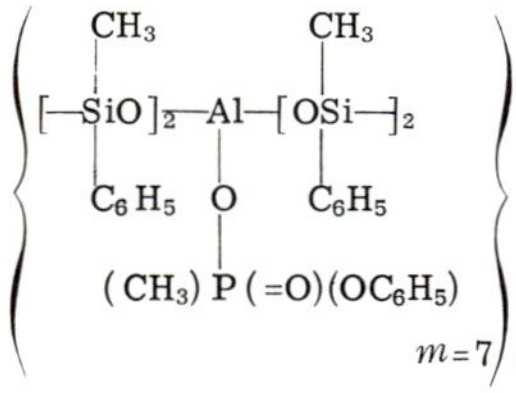	15.30	15.05	+0.25	3.70	3.64	+0.06	4.15	4.17	-0.02
		15.30		+0.25	3.49		+0.15	4.28		+0.11
4	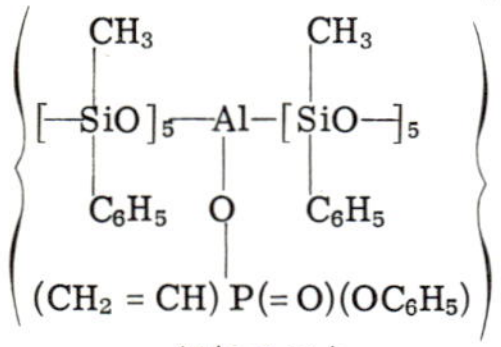	17.42	17.86	-0.44	1.68	1.71	-0.03	2.33	1.96	+0.37
		17.88		+0.02	1.55		-0.16	2.30		+0.34
5	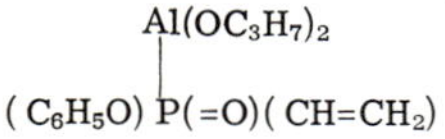	–	–	–	8.52	8.64	-0.12	10.32	9.92	+0.40
					8.58		-0.06	10.01		+0.09
6	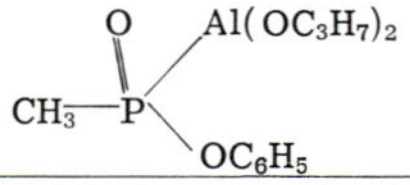	–	–	–	8.50	8.53	-0.03	9.30	9.79	-0.49

EXPERIMENTAL PART

The compound to be analyzed is first tested for solubility in sulfuric acid. If it does not dissolve in sulfuric acid within 10-15 minutes, a weighed sample (40-60 mg) is placed in a microbeaker (capacity 1-1.5 ml) and 1 ml of ether is added. The substance dissolves within a few minutes and the ether is partly evaporated under a hood, leaving a syrupy mass that fills approximately 1/3 of the beaker. The excess ether may also be conveniently removed under suction. For this, the beaker is covered with a funnel which is connected to an aspirator. The beaker with the dissolved sample is placed in a 100-ml Erlenmeyer flask containing 10 ml of sulfuric acid, and the contents are stirred until solution is complete. The clear, thin film formed is completely dissolved. A funnel is placed in the flask to prevent spattering, and the solution is heated until it darkens. The solution is cooled, 0. 5 g of potassium persulfate is added, and the solution is boiled for 10-15 minutes. The addition is repeated another 6-8 times, in 0.5 g batches. In compounds where m equals or exceeds 13, the decomposition is carried out for 30 minutes with 2-3 g of $K_2S_2O_8$. Thirty milliliters of water is carefully added to the cooled solution, and it is boiled for 10-15 minutes. The silicic acid is filtered off, and the silicon is determined acidometrically. [2] The filtrate after separation of the silicon is transferred to a volumetric flask, and aliquot parts are used to determine aluminum complexometrically [1], and phosphorus photometrically in the form of phosphomolybdenum blue. [2]

SUMMARY

A method has been developed for the analysis of difficultly oxidizable organic polymers containing silicon, phosphorus, and aluminum, based upon their oxidation by potassium persulfate in concentrated sulfuric acid.

LITERATURE CITED

[1] A. P. Terent'ev, B. M. Luskina, and S. V. Syavtsillo, *Zhur. Anal. Khim.* **16**, 635 (1961).

[2] B. M. Luskina, A. P. Terent'ev, and S. V. Syavtsillo, *ibid.* **17**, 639 (1962).

[3] B. M. Luskina, A. P. Terent'ev, and S. V. Syavtsillo, *Trudy Komissii Anal. Khim., Akad. Nauk S.S.S.R., Inst. Geokhim. i Anal. Khim.* **13**, 3 (1963).

[4] W. F. Hillebrand, G. E. F. Lundell, et al.: "Laboratory Manual of Inorganic Analysis" [Russian Ed.], Moscow (1957), p. 520.

[5] "Gmelins Handbuch der anorganischen Chemie," 8th Ed., System No. 35, Part B, p. 347 (1934).

[6] E. I. Minsker, M. F. Borisov, and N. F. Orlov, *Vysokomolekul. Soedin.* (*in press*).

[7] E. I. Minsker, M. F. Borisov, et al., *Zhur. Obshch. Khim.* (*in press*).

17

The Microdetermination of Copper in Organic Compounds by Combustion in the Oxygen Flask*

A. P. Terent'ev, S. I. Obtemperanskaya,
and
V. N. Likhosherstova

The attention of analysts has recently been drawn to the Schöniger method, permitting the determination of a number of elements in nonvolatile organic compounds. A sample of the substance is wrapped in a filter paper, attached to a platinum wire, and burned in a flask filled with oxygen. The method was used successfully for the determination of halogen [1], sulfur [2], and phosphorus [3]. It can evidently be used also for the analysis of organic complexes of a number of metals. In this connection we shall describe a technique for the analysis of inner complexes of copper.

A sample containing 4-5 mg copper, weighed on a microbalance into a small glass tube with a long handle, is poured on to the center of an ashless filter. The filter is folded, leaving a free end, and attached to the end of a platinum wire, sealed into a stopper, as indicated in the Schöniger method. [1, 2]

20 ml 0.5 N sulfuric acid is poured into a 500 ml Erlenmeyer flask, and oxygen from a cylinder is passed in for a few seconds. The free end of the filter is ignited, and it is immediately introduced into the

* *Zhur. Anal. Khim.* 15, 748-749 (1960)

flask. The stopper is held in by hand, since it may be forced out. Combustion is complete within a few seconds. The flask is then shaken, a few drops of water are poured on the ground joint of the stopper, and it is left for 15-20 minutes. The stopcock is opened, the stopper is removed and washed with water. The solution from the flask is poured into a 100-ml Erlenmeyer flask and heated to boiling. The platinum wire is ignited for a few seconds in the oxidizing flame of a gas burner, in order to oxidize the small amount of metallic copper formed, and is then lowered into the flask containing the solution and boiled for 1-2 minutes. After all the copper oxide has dissolved, the wire is washed with water and the solution is evaporated to 15 ml.

25% ammonia is added to the cooled solution until a faint blue color appears, then 5 ml 50% acetic acid is poured in, about 1 g potassium iodide is added, and the liberated iodine is titrated with 0.02 N sodium thiosulfate until a light yellow color appears. One ml of 1% starch solution is then added, and titration is continued until the color changes from blue-brown to white-pink.

The error in the method is ±0.2%. The determination takes about 30 minutes. The results of the determination are shown in Table 1.

SUMMARY

A rapid method for determining copper in organic compounds has been developed, based on combustion in an oxygen-filled flask, followed by iodometric titration of the copper. The error is ±0.2%.

TABLE 1

Compound	mg Sample	% Cu Found	% Cu Calc.	% Cu Δ
Salicylaldehyde copper(II) complex	23.091	20.83	20.73	+0.10
$C_{14}H_{10}CuO_4$	22.919	20.61		-0.18
N-Salicylidenemethylamine copper(II)	24.259	19.35	19.15	+0.20
complex $C_{16}H_{16}CuN_2O_2$	24.919	19.28		+0.13
Salicylidenimine copper(II) complex	27.438	21.00	20.90	+0.10
$C_{14}H_{12}CuN_2O_2$	27.010	20.88		-0.02
β-Hydroxynaphthaldehyde copper(II)	23.832	15.63	15.65	-0.02
complex $C_{22}H_{14}CuO_4$	23.450	15.77		+0.12
Benzoylacetone copper(II) complex	14.537	16.44	16.37	+0.07
$C_{20}H_{18}CuO_4$	18.114	16.32		-0.04
Benzoylacetaldehyde copper(II) complex	19.253	17.65	17.78	-0.13
$C_{18}H_{14}CuO_4$	17.388	17.85		+0.07

LITERATURE CITED

[1] W. Schöniger, *Mikrochim. Acta* 1955, 123.
[2] W. Schöniger, *ibid.* 1956, 869.
[3] K. D. Fleischer, et al., *Anal. Chem.* 30, 152 (1958).

18

*The Quantitative Determination of Thioalcohols and Thiophenols with Acrylonitrile**

S. I. Obtemperanskaya, A. P. Terent'ev
and
M. M. Buzlanova

Thioalcohols and thiophenols are usually determined iodometrically [1, 2], argentometrically [3, 4], or by a method based on titration with a solution of copper oleate or copper alkyl phthalate [5, 6]. Of these, the iodometric method is the most accurate.

We have developed a new method for the analysis of thioalcohols and thiophenols, based on the cyanoethylation reaction, which is as accurate as the iodometric method. Thioalcohols and thiophenols react with acrylonitrile as follows:

$$CH_2 = CH\text{-}CN + RSH \rightarrow RS\text{-}CH_2CH_2\text{-}CN \quad [7]$$

We have found that if a sample of a compound containing a mercapto group is treated with excess acrylonitrile (1:3), the reaction goes quantitatively to completion without a catalyst. For thiophenols and lower thioalcohols, the reaction is complete in 1-2 minutes, for higher thioalcohols in one hour. The proposed new method of determination is based on the

* *Vestnik Moskov. Univ., Ser. Mat., Mekh., Astron., Fiz., Khim.* **12**, *No.* 3, 145-147 (1957)

treatment of the thioalcohol or thiophenol sample with a known amount of a dioxane solution of acrylonitrile, and determination of excess acrylonitrile by our previously developed sulfite method. [8] The error in the determination is 0.3%.

The results of the determination of pure thioalcohols and thiophenols are given in Table 1.

REAGENTS

Acrylonitrile (dried with $CaCl_2$ and distilled at atmospheric pressure); dioxane (freed of peroxides by boiling and distillation over sodium); 1 N Na_2SO_3 (240g $Na_2SO_3 \cdot 7H_2O$ in 1 l of boiled, distilled water); 0.1 N HCl; thymolphthalein (0.1 g dye in 100 ml alcohol); Alizarin Yellow (0.1 g dye in 100 ml distilled water).

PROCEDURE

The acrylonitrile solution is prepared by introducing a weighed amount of acrylonitrile (1.1-1.3g) into a 50-ml volumetric flask and filling to the mark with dioxane. A 50-100-mg sample of thiophenol or thioalcohol is placed in a ground-stoppered Erlenmeyer flask, 5 ml acrylonitrile solution is added with a pipette, and the mixture is stirred. After 1-5 minutes (in the case of thiophenols and lower thioalcohols), 20 ml of 1 N Na_2SO_3 is added, and the liberated alkali is titrated with 0.1 N HCl, using a mixed indicator (10 drops thymolphthalein and 2 drops Alizarin Yellow), until a yellow color appears. A blank is run simultaneously. In the case of higher thioalcohols, the addition of sulfite and titration are carried out after an hour. The titer of the acrylonitrile solution is quite stable and does not change over a period of several days.

TABLE 1

Sample No.	Compound	mg Sample	Mole ratio of reagents (RSH:AN)	Reaction time (minutes)	% -SH Calc.	% -SH Found	% -SH Δ
1	Benzenethiol	75.1	1:3.2	1	30.02	30.14	+0.12
		81.8				29.98	0.04
2	*p*-Toluenethiol	72.9	1:2.3	1	26.64	26.74	+0.10
		84.0	1:2.3			26.50	-0.14
3	2-Naphthalenethiol	66.4	1:2.8	1	20.64	20.59	-0.05
		87.1				20.64	0.00
4	α-Toluenethiol	96.6	1:2.5	5	26.63	26.68	+0.05
		103.3	1:2.5	5		26.63	0.00
5	1-Decanethiol	55.7	1:3.8	50	18.97	18.91	-0.06
		62.5	1:3.8	90		19.11	+0.14
6	2-Mercaptopropionic acid	78.0	1:2.8	1	31.16	31.23	+0.07
		55.9	1:3	1		31.07	-0.09
7	2, 3-Dimercaptopropionic acid	54.5	1:2.8	1	23.93	23.84	-0.09
		46.5		1		24.00	+0.07

The calculation is carried out using the formula

$$\% \text{-SH} = \frac{(a-b) \cdot n \cdot 33.08 \cdot 100}{c \cdot 1000}$$

where a = ml of acid required for titration of the blank; b = ml of acid required for titration of the sample; c = sample weight (grams); n = normality of the acid.

SUMMARY

A new method for the analysis of thioalcohols and thiophenols, based on the cyanoethylation reaction, is proposed. The accuracy of the method is 0.2-0.3%.

It is shown that thioalcohols and thiophenols react quantitatively with acrylonitrile without a catalyst, if a large excess of acrylonitrile is used.

LITERATURE CITED

[1] P. Klason and T. Carlson, *Chem. Ber.* **39**, 738 (1906).
[2] J. W. Kimball, R. L. Kramer, and E. E. Reid, *J. Am. Chem. Soc.* **43**, 1199 (1921).
[3] P. Borgstrom and E. E. Reid, *Ind. Eng. Chem., Anal. Ed.* **1**, 186 (1929).
[4] W. M. Malisoff and C. E. Anding, Jr., *ibid.* **7**, 86 (1935).
[5] G. R. Bond, Jr., *ibid.* **5**, 257 (1933).
[6] E. Turk and E. E. Reid, *ibid.* **17**, 713 (1945).
[7] O. Bayer, *Angew. Chem.* **61**, 229 (1949).
[8] A. P. Terent'ev, S. I. Obtemperanskaya, and M. M. Buzlanova, *Vestnik Moskov. Univ., Ser. Mat., Mekh., Astron., Fiz., Khim.* **11**, *No.* 1, 187 (1956).

19

*The Quantitative Determination of Monohydric Alcohols with Acrylonitrile**

S. I. Obtemperanskaya, A. P. Terent'ev,
and
M. M. Buzlanova

In a previously published paper [1], we showed that acrylonitrile is a convenient reagent for the quantitative determination of thioalcohols and thiophenols. The reaction between acrylonitrile, thioalcohols, and thiophenols is quantitative without a catalyst. A further study of the possibility of using acrylonitrile in organic analysis showed that this reagent can be used for the quantitative determination of alcohols, if the reaction is carried out in alkaline medium. Acrylonitrile reacts with alcohols according to the equation [2]:

$$ROH + CH_2{=}CH{-}CN \rightarrow ROCH_2CH_2CN.$$

By introducing into the reaction a known excess of acrylonitrile, dissolved in dioxane, and determining this excess by our previously proposed sulfite method [3, 4], primary and secondary alcohols and alcohols with some other functional groups can be determined. In polyhydric alcohols (ethylene glycol), only one hydroxy group reacts quantitatively in the cold.

**Zhur. Anal. Khim.* 16, 372-374 (1961)

For primary alcohols, the reaction is complete in 1-1½ hours at room temperature, or 15-30 minutes at 35-40°; for secondary alcohols it takes 3-6 hours at room temperature, and 1-1½ hours at 35-40°. A necessary condition is the careful purification of the dioxane and acrylonitrile, to remove peroxides and other impurities. Water interferes with the determination, since it reacts with acrylonitrile in the presence of alkali.

The acrylonitrile is shaken with NaOH solution, dried first with KOH, then with $MgSO_4$, and distilled. The dioxane is refluxed for several hours with finely cut metallic sodium and then distilled from it.

The results for the determination of alcohols are shown in Table 1.

PROCEDURE

A weighed sample (about 1 millimole) of alcohol is introduced into a 25 ml volumetric flask, and 5 ml dioxane and 0.02-0.03 g of finely ground KOH are added. The flask is closed with a stopper, carefully shaken until the alkali is completely dissolved, and made up to the mark with dioxane. 1 ml of the prepared solution is introduced into a ground-stoppered Erlenmeyer flask, 2 ml of 1 N acrylonitrile (72 g acrylonitrile in 1 l dioxane) is added, the flask is stoppered and let stand (for primary alcohols: 1-1½ hours at room temperature or 15-30 minutes at 35-40°; for secondary alcohols: 3-6 hours at room temperature or 1-1½ hours at 35-40°).

After the reaction is complete, the excess of acrylonitrile is determined: 25 ml of 1 M Na_2SO_3 (240g of $Na_2SO_3 \cdot 7H_2O$ per liter of boiled, distilled water) is poured into the flask, 5-7 drops of mixed thymolphthalein-Alizarin Yellow indicator (prepared by

TABLE 1

Compound	Time (hrs)	Temp. (°C)	% OH Calc.	% OH Found
Methyl alcohol	0.75	Room	53.06	53.10
	1.00	Room		52.98
	0.33	30-35		53.10
	0.33	30-35		53.04
Ethyl alcohol	1.75	Room	36.90	36.88
	1.75	Room		36.99
Propyl alcohol	1.00	Room	28.29	28.25
	1.25	Room		28.32
Butyl alcohol	1.50	Room	22.94	22.92
	1.50	Room		22.95
	0.25	45		22.92
	0.33	45		22.95
Pentyl alcohol	0.66	Room	19.28	19.28
	0.66	Room		19.20
Heptyl alcohol	1.33	Room	14.63	14.62
	1.33	Room		14.57
	0.58	35-40		14.62
	0.58	35-40		14.67
Allyl alcohol	1.00	Room	29.27	29.32
	1.00	Room		29.19
Benzyl alcohol	0.25	Room	15.73	15.71
	0.41	Room		15.76
Isobutyl alcohol	0.66	Room	22.97	22.98
	0.66	Room		22.88
Isopentyl alcohol	0.66	Room	19.28	19.29
				19.35

TABLE 1 (cont'd)

Compound	Time (hrs)	Temp. (°C)	%OH Calc.	%OH Found
Cyclohexane-methanol	0.50	Room	14.89	14.91
	0.50	Room		14.88
Phenethyl alcohol	0.50	Room	13.91	13.93
	0.50	Room		13.85
Phenylpropyl alcohol	0.50	Room	12.48	12.54
				12.48
Isopropyl alcohol	0.75	Room	28.29	28.32
	5.50	Room		28.32
Cyclohexanol	2.50	Room	16.97	17.02
	2.50	Room		16.98
	1.00	35-40		16.93
	1.00	35-40		16.93
sec-Butyl alcohol	1.00	35-40	22.97	22.97
	1.00	35-40		22.83
4-Heptanol	1. 50	35-40	14.63	14.63
				14.60
Ethylene cyanohydrin	3. 00	Room	23.92	23.91
				23.93
2-Ethoxyethanol	0.50	Room	18.87	18.85
	0.50	Room		18.87

mixing 30 ml of 0.1% alcoholic thymolphthalein and 15 ml of 0.1% aqueous Alizarin Yellow) is added, and the liberated alkali is titrated with 0.1 N HCl until the blue color changes to yellow.

A blank is carried out simultaneously: 2 ml acrylonitrile solution is added to 25 ml Na_2SO_3 solution, and after 5 minutes 1 ml of the prepared solution of the alcohol under study is added. The mixture is titrated with 0.1 N acid, using the same indicator.

The calculation is carried out according to the formula

$$\% \text{ OH} = \frac{(a-b) \cdot N \cdot 17}{c \cdot 10}$$

where a, b = ml of acid used in the titration and in the blank, c = weight of sample, N = normality of the acid.

LITERATURE CITED

[1] S. I. Obtemperanskaya, A. P. Terent'ev, and M. M. Buzlanova, *Vestnik Moskov. Univ., Ser. Mat., Mekh., Astron., Fiz., Khim.* 12, No. 3, 145 (1957).

[2] J. H. MacGregor and C. Pugh, *J. Chem. Soc.* 1945, 535.

[3] A. P. Terent'ev and S. I. Obtemperanskaya, *Zhur. Anal. Khim.* 11, 638 (1956).

[4] A. P. Terent'ev, S. I. Obtemperanskaya, and M. M. Buzlanova, *Vestnik Moskov. Univ., Ser. Mat., Mekh., Astron., Fiz., Khim.* 11, No. 1, 187 (1956).

20

A New Apparatus for the Microdetermination of Active Hydrogen*

M. Souček

Three procedures have been developed for the determination of active hydrogen by the Zerewitinoff method [1], differing in the measurement of the gas formed in the reaction. The oldest and most widely used is the volumetric determination. [2-6] However, this method has a number of disadvantages: inaccurate reading of volumes of the gas burette; the need for a blank, with strict maintenance of the same experimental conditions; lengthy thermostating and, therefore, lengthy time required for the determination. In 1940, a gravimetric determination of active hydrogen was described [7], in which the CH_4 formed is burned, and the water and CO_2 formed are weighed. This method in itself is very complex, and is accompanied by a number of experimental errors. In recent years, a manometric determination has been developed [8, 9], in which the amount of gas formed is calculated from the measured pressure increase and the known volume of the apparatus. However, this method requires a troublesome and time-consuming calibration of the apparatus, and a very accurate maintenance of constant temperature.

* *Collection Czechoslov Chem. Communs.* **23**, 554-557 (1958)

In order to overcome the difficulties cited, we have constructed in our laboratory an apparatus operating on a new principle. The gas formed in the reaction of Grignard solution or $LiAlH_4$ with a compound containing active hydrogen forces an equivalent volume of measuring liquid from the lower into the upper manometer container. By a gradual pressure decrease, which is achieved by drawing off the measuring liquid from a burette incorporated in the system, the liquid levels in the manometer are equalized. The volume of liquid drawn off is equal to the volume of gas formed.

EXPERIMENTAL

APPARATUS

The whole apparatus (Fig. 1) is made of soft glass with a high coefficient of expansion. It consists of three closely connected parts: reaction vessel with head, manometer, and burette. The reaction vessel R is a flat-bottomed 12-ml cylindrical flask, whose upper part is a male No. 29 standard-taper ground joint. The lower part of the head of the reaction vessel is a female No. 29 standard-taper ground joint, whose upper part carries three capillaries a, b, c (3 mm i. d.). Nitrogen is passed into the system through the capillary c with stopcock k_2. This capillary passes into the interior of the reaction flask and is bent in such a way that the reaction basket containing the analytical sample can be hung on its hooked end. The glass basket K (5 mm high, 9 mm wide) has an iron rod sealed into it, thus permitting magnetic stirring with the whole basket. Capillary a is sealed to the manometer; the capillary b, with a W3 female

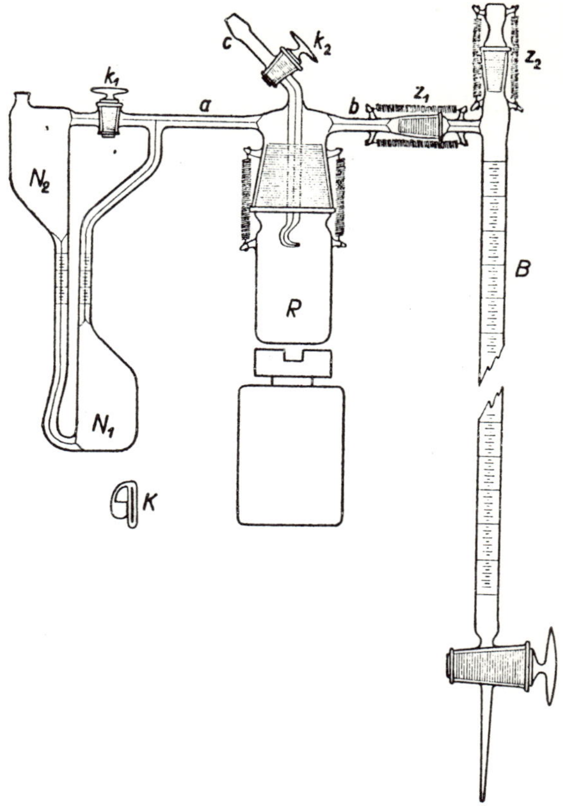

Fig. 1 Apparatus for the determination of active hydrogen.

standard-taper ground joint, connects the head of the reaction vessel to the burette. The manometer has two equalization chambers N_1 and N_2, whose position and construction can be seen from Fig. 1. Chamber N_1 is connected to the reaction vessel through the capillary arm and capillary a; N_2 is open to the atmosphere. Stopcock k_1 connects N_2 to capillary a. The manometer arms are graduated to simplify the reading of the liquid levels in the manometer. The burette B (capacity 10 ml, with 0.02 ml graduations) is closed with the ground stopper Z_2 and is connected to the head of the reaction vessel through the side tube with male ground joint Z_1.

REAGENTS

Dibutyl ether was dried for 10 days over sodium and 24 hours over $LiAlH_4$, decanted, and distilled; b. p. 136-137°C.* Diamyl ether was heated with sodium for 48 hours and distilled; b. p. 172-174°C. *

The solution of CH_3MgI in diamyl ether was prepared in the usual manner [10]. The $LiAlH_4$ solution was prepared by shaking $LiAlH_4$ with dibutyl ether for 6 hours. The turbidity was removed by centrifugation, and the reagent was charged into ampoules under nitrogen and sealed off. The solution was not noticeably changed after three months.

"Medicinal" nitrogen was dried by passage through a $CaCl_2$ drying tower, a U-tube filled with Anhydrone, and a spiral wash bottle filled with Grignard solution or $LiAlH_4$ solution. Prepurified nitrogen (from Ostrava Nitrogen Works) may be used instead of the "medicinal" nitrogen. Nitrogen with a higher oxygen content must be passed through a glowing tube filled with copper wire before drying. [11]

---

*(Ed. Note): These boiling points do not agree with the values given in "Handbook of Chemistry and Physics", 43rd Edition (1961-1962): dibutyl ether 142°, di-sec-butyl ether 121°, diisobutyl ether 122.5°, diamyl ether 190°, diisoamyl ether 172.5-173°.

PROCEDURE

The whole apparatus is rinsed successively with dilute HCl (1:5), distilled water, ethanol, and diethyl ether, and is dried at 130°C. in a drying oven. The ground joints, while still hot, are lubricated with solid Ramsay grease, and the apparatus is assembled. A short test tube with Anhydrone is attached instead of the reaction vessel. The manometer and burette, while still hot, are filled with dibutyl or diamyl ether. With stopcock k_1 and stopper Z_2 open, the apparatus is cooled with dry nitrogen. In the meantime, the analytical sample is weighed into the dry basket, and 2 ml of solvent and 1.5 ml of reagent are pipetted into the reaction flask with a carefully dried pipette. (For a microdetermination, 0.5 ml of reagent is used). The reaction vessel is closed with the ground head, and the basket with the weighed sample is placed in a desiccator filled with Anhydrone. After the apparatus has cooled, the burette is closed with the stopper Z_2, the test tube with the desiccant is removed, the basket with the sample is hung on the bent end of the capillary, and the reaction vessel is attached. In order to obtain complete temperature equalization in the apparatus, nitrogen is passed through the apparatus for another 15 minutes. The stopcocks k_1 and k_2 are then closed. If there is a reduced pressure in the system (the liquid in the manometer rises in the direction of capillary a) this means that the apparatus had not cooled sufficiently; if the liquid rises in the direction of the equalizing vessel N_2, there is an excess pressure in the system, and it is probable that the apparatus has not been properly dried.

After equalizing the liquid levels in the manometer, the determination can be started. The basket with the sample is dropped into the reagent by means of a magnet, and the magnetic stirring is started. For most compounds, the reaction is complete after 15 minutes, and the temperatures are equalized. After turning off the stirrer, liquid is drawn off from the burette until the liquid levels in the manometer are equalized. In order to check whether the pressure in the system remains unchanged, one waits another 5 minutes. The volume of gas evolved in the reaction is read off from the burette. For subsequent determinations, the reaction vessel and basket are merely replaced by new ones.

The calculation is carried out in the usual manner. When using solvents whose vapor pressure at the laboratory temperature is greater than 10 mm Hg, this value must be subtracted from the barometric pressure.

RESULTS AND DISCUSSION

The measurements carried out by means of the apparatus described are summarized in Table 1. The reproducibility of the results shows that this method is suitable for the semimicro- and microdetermination of active hydrogen; the values obtained are in good agreement with literature values (cf., e.g., [6], [12], [13].)

The new apparatus has a number of advantages over the types used hitherto. It is sufficient to maintain the temperature constant within ± 1°C. in the vicinity of the apparatus, since, in view of the small volume of the apparatus, the volume change produced by this fluctuation may almost be neglected, and does not exceed the limits of experimental error when the apparatus is made of a soft glass with a high coefficient of expansion. The principle of magnetic stirring

TABLE 1

mg Sample	Reagent	Gas volume (ml, at 0°C/760 mm) Calc. *	Found	Number of active H
		Benzoic Acid		
15.43	$LiAlH_4$	2.83	2.92	1.03
14.70	$LiAlH_4$	2.70	2.86	1.06
10.32	$LiAlH_4$	1.89	1.93	1.02
3.686	$LiAlH_4$	0.68	0.70	1.03
5.321	$LiAlH_4$	0.98	1.02	1.05
4.953	$LiAlH_4$	0.80	0.82	1.03
13.83	CH_3MgI	2.54	2.66	1.05
11.02	CH_3MgI	2.02	2.10	1.04
11.37	CH_3MgI	2.08	2.14	1.03
		o-Nitrophenol		
12.89	$LiAlH_4$	2.08	2.27	1.09
20.32	$LiAlH_4$	3.74	4.03	1.08
16.14	CH_3MgI	2.60	2.65	1.02
11.08	CH_3MgI	1.79	1.84	1.03
3.128	CH_3MgI	0.50	0.51	1.02
3.592	CH_3MgI	0.58	0.60	1.04

* In the calculation, the value 22.36 was used for 1 mole of gas for CH_4 (cf., e.g., [10]), and the value 22.41 for H_2.

[2] and weighing the sample into the basket have the advantage that one can work without a blank, the value of which in other methods is up to 10% of the result. However, there is the danger that compounds that are only slightly soluble in the solvent used will not react

quantitatively. For this reason, stirring must be intensive, and the basket must be wide and low.

For compounds that dissolve only on heating, the apparatus can be modified by replacing the proposed reaction flask by a Zerewitinoff vessel.

LITERATURE CITED

[1] T. Zerewitinoff, *Chem. Ber.* **40**, 2023 (1907).

[2] W. Fuchs, N. H. Ishler, and A. G. Sandhoff, *Ind. Eng. Chem., Anal. Ed.* **12**, 507 (1940).

[3] F. A. Hochstein, *J. Am. Chem. Soc.* **71**, 305 (1949).

[4] E. P. Kohler and N. K. Richtmyer, *ibid.* **52**, 3736 (1930).

[5] S. Siggia: "Quantitative Organic Analysis via Functional Groups", New York: John Wiley (1948), pp. 41 ff.

[6] A. Soltys, *Mikrochemie* **20**, 107 (1936).

[7] R. N. Evans, J. E. Davenport, and A. J. Revukas, *Ind. Eng. Chem.*, *Anal. Ed.* **12**, 301 (1940).

[8] J. A. Krynitsky, J. E. Johnson, and H. W. Carhart, *Anal. Chem.* **20**, 311 (1948).

[9] J. A. Krynitsky, J. E. Johnson, and H. W. Carhart, *J. Am. Chem. Soc.* **70**, 486 (1948).

[10] M. Jureček: "Organická analysa" [Organic Analysis], Prague (1950), p. 270.

[11] F. Pregl and H. Roth: "Quantitative organische Mikroanalyse", 4th ed., Berlin: Springer (1935), p. 197.

[12] H. E. Zaugg and B. W. Horrom, *Anal. Chem.* **20**, 1026 (1948).

[13] I. M. McAlpine and P. A. Ongley, *ibid.* **27**, 55 (1955).

21

*The Semimicrodetermination of Active Hydrogen and Quantitative Hydrogenation**

B. Buděšínský

The numerous apparatuses constructed for gas-volumetric semimicro- or microdeterminations in organic chemistry differ in the permissible service load, in the measurement of gas volume, in the type of mixing of the materials in the closed system of the apparatus, in the nature of thermostating, in the size of the dead spaces, and in the versatility of their application. Details will be found in the corresponding review article. [1] The accuracy of the determinations is generally not high; the errors are sometimes as high as ± 10%.

Very accurate results in the determination of active hydrogen can be obtained by means of the Souček apparatus. [2] The disadvantage of this apparatus is that it can be used to determine active hydrogen only in compounds that have low volatility and good solubility in dibutyl or diamyl ether.

Since, in organic analysis, gas-volumetric determinations are not as frequent as, for example, elemental determinations (C, H, etc.), we concerned ourselves with the design of a universal apparatus suitable for this lower service load and operating with the greatest possible accuracy.

**Collection Czechoslov. Chem. Communs.* 24, 2948-53 (1959)

In constructing the apparatus, we adopted the manometric measurement of gas volume of the Součka apparatus, and supplemented it with an arbitrary and sufficiently fine adjustment of the mercury level in the gas burette and with a dropping funnel reminiscent of the dropping burette of the earlier, much more complicated apparatuses (e.g., that of Soltys [3].) We have used the apparatus successfully for the determination of active hydrogen and for quantitative hydrogenation.

EXPERIMENTAL

REAGENTS

The solutions of $LiAlH_4$ in dibutyl ether and CH_3MgBr in diamyl ether, and the solvents xylene, anisole, dioxane, and pyridine were prepared in the usual manner. For individual determinations of active hydrogen, 5-ml samples of the solutions were sealed into ampoules.

In hydrogenation, we worked with a solution of $PdCl_2$ (60 mg Pd/ml), with Raney nickel [4], or with PtO_2. [5] Ethanol and acetic acid were used as solvents, their quality being checked by blanks.

Prepurified nitrogen (from Ostrava Nitrogen Works) was washed with alkaline pyrogallol solution and dried with $CaCl_2$ and P_2O_5. Hydrogen was prepared from pure zinc and HCl in a Kipp generator, and was washed with $KMnO_4$ solution containing 20% H_2SO_4 and with KOH solution (2:1).

The quality of the organic compounds used was checked by the usual methods of elemental organic analysis.

The apparatus (Fig. 1) consists of a reaction

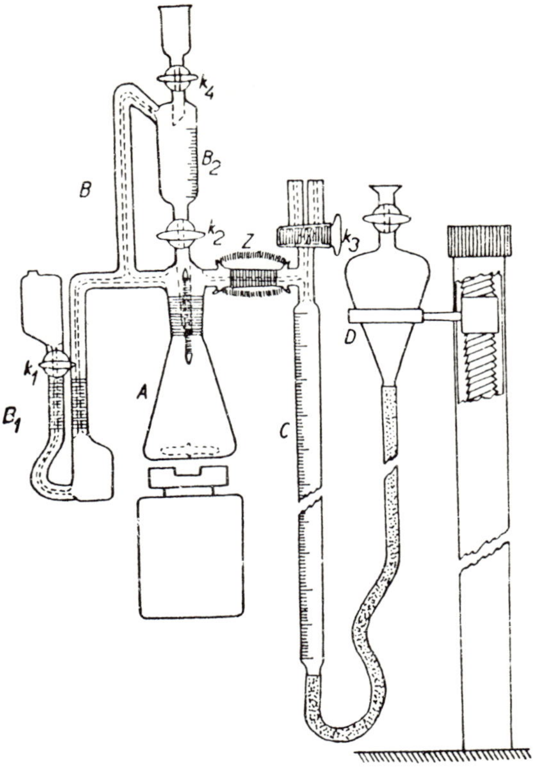

Fig. 1 Apparatus for determination of active hydrogen and quantitative hydrogenation.

flask A (about 15 ml) with magnetic stirrer. Another part B of the apparatus has a manometer B_1, filled with xylene, dibutyl or diamyl ether, on the left side. The i. d. of the capillaries is about 1.5 mm. Accurate equalization of the liquid levels in the manometer is facilitated by a finely divided scale, or by a piece of graph paper placed behind the manometer. By opening or closing the stopcock K_1, the manometer can be connected to or disconnected from the rest of the apparatus. The central part of section B is a graduated dropping funnel B_2 (about 7 ml) with stopcock k_2, through which the required amount of solution is introduced into the reaction flask A in the closed system of the apparatus. For this purpose, the dropping funnel is also connected with the system by means of a capillary by-pass at its top. The dropping funnel B_2 is filled through the small upper funnel. The stopcock k_4 connecting the two funnels also serves to purge the whole apparatus. On the right side of section B is the ground joint Z, attached to a 25-ml gas burette with 0.05-ml graduations. At the upper end of the burette there is a two-way stopcock k_3 (e. g., for passing in hydrogen or nitrogen); at the lower end it is connected by means of a rubber tube to the leveling bulb and pressure regulator D. The pressure regulator can be raised or lowered by means of a screw with a steep pitch. Mercury is used as the connecting liquid between the gas burette and the leveling bulb. The ground joints of the apparatus are lubricated with Ramsay grease. To improve temperature constancy, it is desirable to place the whole apparatus into a closable box (e. g., of Plexiglas).

DETERMINATION OF ACTIVE HYDROGEN

The volatility of the compound and its solubility in xylene, anisole, dioxane, or pyridine are first determined. If the sample is nonvolatile (b.p. above 120°C) and dissolves readily in dibutyl or diamyl ether, the determination is carried out without preliminary solution of the sample. Compounds that are volatile (b. p. below 120°C) or poorly soluble in dibutyl or diamyl ether are always dissolved in a suitable solvent.

About 0.8 milliequivalents of the compound (about 40 mg, if the composition is unknown) is then accurately weighed into the reaction flask, and dissolved in 5 ml of a suitable solvent, if necessary. Solution in the reaction flask is brought about by means of the magnetic stirrer. The ampoule of $LiAlH_4$ or CH_3MgBr solution is then opened, the contents are poured through the small funnel into the dry dropping funnel B_2, and the nitrogen stream is turned on by means of the two-way stopcock k_3. The mercury column in the burette is set at the upper zero mark and, with stopcock k_4 open (the manometer stopcock k_1 is closed), nitrogen is passed through the apparatus for 5 minutes. First stopcock k_4, then the nitrogen stopcock are closed, and the apparatus is thermostated for at least 15 minutes. After the temperature has become constant, the equalized levels of the manometer liquid should not change further after opening the manometer stopcock and after equalizing the pressure by lowering the leveling bulb. If the gas volume in the apparatus still increases after 30 minutes, with the external temperature constant, this indicates that the apparatus has not been adequately cleaned and dried. If the gas volume does not change further, the height of the mercury column in the burette (volume V) is read off, after equalizing the manometer level. After closing the manometer stopcock, the mercury level is lowered to about the 20 ml mark by lowering the leveling bulb; then, after waiting for a few minutes and checking the gas-tightness of the apparatus by readings of the manometer level, the stirrer is turned on, and the contents of the dropping funnel are added to the reaction flask by opening stopcock k_2. After the reaction is complete, the apparatus is again thermostated for 15 minutes, and the position of the leveling bulb is then adjusted so that the mercury level in the bulb is about

TABLE 1

Determination of Active Hydrogen

Compound	mg Sample	Reagent	% active H Calc.	% active H Found
Benzoic acid	95.13	$LiAlH_4$	0.827	0.823
	80.52			0.827
				0.830
	98.16	CH_3MgI		0.831
	96.50			0.822
	92.17			0.823
Dimethyl phenyl-malonate	151.31	CH_3MgI	0.485	0.488
				0.482
Ethyl cyclohexyl-phenylglycolate	198.30	$LiAlH_4$	0.384	0.386
				0.382
Diphenylcarbinol	145.13	$LiAlH_4$	0.548	0.545
	142.48			0.551
Methanol*	25.12	$LiAlH_4$	3.12	3.09
	24.88			3.10
	26.50	CH_3MgI		3.15
	25.76			3.14
Ethanol*	35.11	$LiAlH_4$	2.17	2.19
	36.04			2.16
	36.37	CH_3MgI		2.15
	34.15			2.18

*Dissolved in 5 ml anhydrous xylene

1 cm above the mercury level in the gas burette. The manometer stopcock is then opened, and the liquid level in the manometer is equalized by fine adjustment of the leveling bulb. The mercury level in the burette (volume V') is again read only after the manometer level has remained unchanged for several minutes.

If the sample has been dissolved in a solvent, a blank must be carried out with the same solvent. The results are calculated by means of equation (I). The results of a series of determinations are summarized in Table 1.

DETERMINATION BY QUANTITATIVE HYDROGENATION

1 ml of $PdCl_2$ solution and about 50 mg active charcoal or 2 ml suspended Raney nickel or about 60 mg PtO_2 are placed in the reaction flask and made up to 5 ml with a suitable solvent (ethanol or cyclohexanol for $PdCl_2$ or Raney nickel; glacial acetic acid for PtO_2). The magnetic stirrer is turned on, and the apparatus is connected. A weighed sample (corresponding to about 0.5 millimoles of hydrogen) of the compound being hydrogenated is dissolved in 5 ml of the suitable solvent and the solution is placed in the dropping funnel. If the composition of the sample is unknown, about 30 mg is used. Hydrogen is passed through the apparatus for 10 minutes in the manner described above, the burette is filled with 20-25 ml hydrogen with the manometer stopcock closed, the hydrogen is turned off, and the storage bulb is raised as high as possible. Under the excess pressure thus produced, the catalyst is saturated with hydrogen, with magnetic stirring. The apparatus is thermostated at the same time. As soon as the gas volume stops decreasing and the mercury level no longer changes, i.e., after about 5-10 minutes, the saturation of the catalyst and thermostating are complete. The gas volume in the burette (volume V') is read off accurately as described above. If the mercury level in the burette is above the 20-ml mark, the required amount of hydrogen is added, the apparatus is again thermostated, and then the volume V' is read off. Then the actual hydrogenation can be started: The

manometer stopcock is closed, the storage bulb is raised to the highest position, and the solution of the compound to be hydrogenated is admitted into the reaction flask by opening stopcock k_2. The hydrogenation is carried out with magnetic stirring. With Raney nickel or palladium, the volume is read off after 15 minutes; with platinum after 5 minutes. After reading off the exact gas volume, the leveling bulb is again raised to the highest position. As a rule, hydrogenation is carried out for 2 hours with Raney nickel, and for about 45 minutes with platinum in glacial acetic acid. However, the hydrogenation time depends on the nature of the compound being hydrogenated, so that a longer reaction time is not excluded. The results obtained are best shown graphically.

$$\%\text{H (active)} = \frac{1.6028 \cdot E(p-m)(V' - V - s) f}{(273 + t) \cdot n} \qquad \text{(I)}$$

The calculation is carried out by means of equation (I), in which V' and V are the observed volumes (ml); p the measured barometric pressure (mm Hg); m the vapor pressure of the solvent used; t the temperature in °C, which must not fluctuate by more than 1°C during the determination; n the sample weight in mg; and s the blank value in the determination of active hydrogen (for hydrogenation, $s = 0$ in the case of a suitable solvent). f is a correction factor of the apparatus for a given solvent in the quantitative hydrogenation, since the solution of sample does not flow quantitatively from the dropping funnel into the reaction flask; $f = w/w'$, where w is the amount of solvent placed in the dropping funnel and w' is the amount leaving it (the f values vary from 1.02-1.04, and are determined by weighing the solvent). In the determination of active hydrogen, f is evidently 1. The equivalent values E are given in Table 2. The results

TABLE 2

Equivalent Values *E*

Group	Active H	C=C	C≡C	NO_2
E	1.008	24.02	12.01	15.34

obtained in quantitative hydrogenation by means of the apparatus described are given in Table 3.

DISCUSSION

The apparatus described has two advantages over the previous apparatuses. It is universally applicable for determination of active hydrogen and quantitative hydrogenation, and the very accurate results obtained in the determination of active hydrogen are equal to those obtained with the Souček apparatus. The determination of active hydrogen in compounds that dissolve readily in dibutyl or diamyl ether can be carried out with this apparatus (in a manner similar to that with the Souček apparatus) without a preliminary blank. The apparatus described has several other advantages over that of Souček: besides its universal applicability, the change in gas volume is completely reversible, which is important particularly in thermostating. The dropping-funnel system is better than the basket-stirrer arrangement [2]; firstly, because of its manifold applicability and secondly because the basket stirrer breaks very easily during the actual determination. A disadvantage of both the present and the Souček apparatus is the large number of ground joints and stopcocks (a total of 5 in the apparatus), whose poor quality causes a continual leak in the apparatus.

In the described determination of active hydrogen,

TABLE 3

Determination of Some Functional Groups by Quantitative Hydrogenation.

Compound	mg Sample	Catalyst	Functional group	% Functional group Calc.	% Functional group Found
Nitrobenzene	19.82	Ni	NO_2	37.4	36.8
	19.82	Pd			37.6
o-Nitrophenol	23.16	Ni	NO_2	33.1	33.8
	23.16	Pd			32.5
Oleic acid	31.30	Pd	C=C	8.5	8.9
	32.14	Pt			8.1
Styrylacetic acid	29.33	Pd	C=C	14.8	14.9
Allyl alcohol	28.55	Pd	C=C	41.4	41.0
Dimethylethynylcarbinol	21.01	Ni	C≡C	29.7	29.4
2-Butyn-1-ol	17.48	Ni	C≡C	34.3	34.5

the sample weight often exceeds the limits of the semimicro method (Table 1). If the described gas burette is replaced by a 10-ml microburette with 0.02-ml graduations, active hydrogen can be determined entirely on a semimicro scale. The errors in the determination are approximately doubled by this change.

By means of the described apparatus, a determination of the carbonyl group by the Strache-Falkenhausen method [6] would also be possible. However, since a better potentiometric modification of this method is known [7], the gas-volumetric determination has lost its previous significance.

By inserting an absorption vessel for nitrogen oxides between the reaction flask and the gas burette, it would also be possible to use the described apparatus for the determination of primary amino groups by the Van Slyke method [8].

SUMMARY

An improved universal apparatus for gas-volumetric determinations is described. The apparatus has been used successfully for the semimicrodetermination of active hydrogen by means of $LiAlH_4$ or CH_3MgI, and for quantitative hydrogenation on palladium, platinum, or Raney nickel.

LITERATURE CITED

[1] G. F. Wright, in "Organic Analysis", ed. J. Mitchell, Jr., et al., New York: Interscience Publishers, Inc. (1953 ff.), vol. I, p. 155; T. Higuchi, *loc. cit.*, vol. II, p. 131.

[2] M. Souček, *Chem. Listy* **50**, 323 (1956); *Collection Czech. Chem. Communs.* **23**, 554 (1958).

[3] A. Soltys, *Mikrochemie* **20**, 107 (1936).

[4] A. A. Pavlic and H. Adkins, *J. Am. Chem. Soc.* **68**, 1471 (1946).

[5] V. Voorhees and R. Adams, *ibid.* **44**, 1397 (1922); R. Adams and R. L. Shriner, *ibid.* **45**, 2171 (1923).

[6] F. v. Falkenhausen, *Z. Anal. Chem.* **99**, 241 (1935).

[7] B. Buděšínský, *Chem. Listy* **52**, 2292 (1958); *Collection Czech. Chem. Communs.* (in press).

[8] D. D. Van Slyke, *Chem. Ber.* **44**, 1684 (1911).

22

*A Universal Apparatus for Gasometric Microdetermination**

B. Buděšínský

A number of apparatuses for gasometric determinations have been described. [1-5] Past development shows that the basic requirements for an ideal apparatus can be summarized in the following four points:

1. Minimum dead space.
2. Minimum number of ground joints.
3. Good thermostating.
4. Operation of apparatus independent of external pressure.

No apparatus constructed so far has satisfactorily met all of these requirements. In searching for such an apparatus, we soon recognized that the classical volumetric or manometric principle of determination must be abandoned in order to satisfy the above four requirements. The solution of the problem lies only in a determination that can be carried out in such a manner that volume and pressure are variable.

**Mikrochim. Acta* 1961, 811-16

EXPERIMENTAL PART

DESCRIPTION OF THE APPARATUS AND ITS FUNCTIONS

The requirements stated above are met by the apparatus shown in Fig. 1.

This consists of a reaction space A (volume V_a), a measuring capillary B (diameter d), and a compression space C (volume $V_č$). The reaction space is provided with inlet stopcock H_1 and outlet stopcock H_2 for passage of gases, and with a sample boat S with ground joint for introducing the sample. At the bottom of the reaction space there is a magnetic stirrer. A millimeter scale is attached to the measuring capillary. The right arm of the capillary is filled with a colored liquid (e.g., dibutyl phthalate with methyl violet). The whole apparatus is so built into a Plexiglas jacket that stopcocks H_1 and H_2 are accessible from above, and the boat S from the side. The ground joint of the boat is secured by steel springs attached to the Plexiglas vessel or by strong rubber bands. A constant temperature is maintained by passing water from an ultrathermostat through the jacket.

The change in the amount of gas in reaction space A can be measured from the height of the liquid column in the measuring capillary. The following relationship holds true for the variables of state of the gas in reaction space A:

$$\Delta n = \pm \frac{(h - h_0)}{81.4\ T} \left(1 + \frac{0.0814\, T L_T v_L}{v_A} \right) \cdot \left[\frac{1.316\, p_s q (v_A + v_C)}{v_C - q(h_0 + h - 2h_s)} + 2 \rho v_A \right] \quad (1)$$

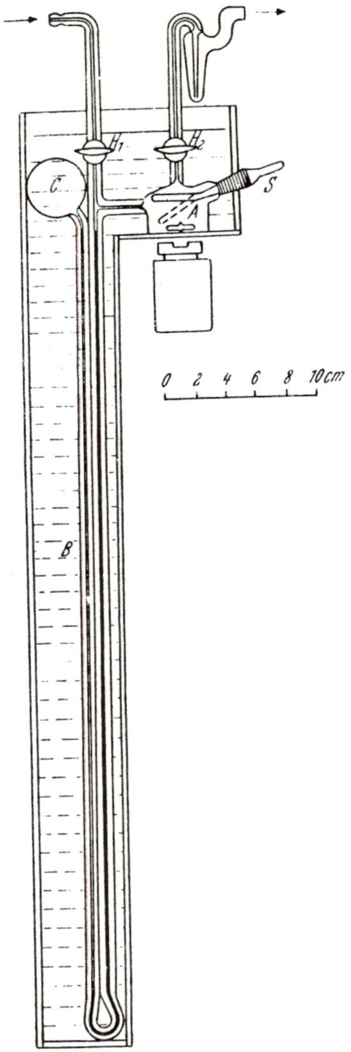

Figure 1

where h_0 is the initial height of the liquid column in the right arm of the measuring capillary, h is the final height, and h_s is the so-called standard height (all in cm), i. e., the height that corresponds to the external pressure p_s (in torr) when the reaction space is opened. T is the absolute temperature, q the cross-section of the measuring capillary (in sq cm), ρ the density of the measuring liquid, L_T the solubility of the gas (in mmoles/ml) in the reaction solution at temperature T, and V_L the volume of the reaction solution (all volumes in ml). The first term in the square brackets in equation (1) corresponds to the pressure change, and is usually several times greater than the second term, which corresponds to the volume change.

The accuracy of determination that can be obtained with this apparatus depends on several factors. First, as shown by equation (1), the difference in height of the liquid column in the measuring capillary depends on the diameter of the capillary. Good temperature constancy and a tightness of the apparatus satisfactory at an excess pressure of about 0.05 to 0.06 atm. are also extremely important.

If there is a permissible error Δh in reading the measuring-liquid level, the useful diameter d of the capillary will be given by the following formula:

$$|v_A - v_C| = \frac{68.4\ \pi d^2\ \Delta h}{E}, \tag{2}$$

where E is the accuracy with which the temperature is maintained constant.

The compressibility condition is expressed by the following equation:

$$v_C = \frac{\pi\ d^2 p_s (h_{max} - h_{min})}{4\ \Delta p} \tag{3}$$

where Δp is the corresponding excess pressure.

Equations (1) to (3) also give the construction elements of the apparatus. Some examples of favorable combinations of the construction elements are given in Table 1.

TABLE 1

Construction Parameters of the Apparatus for $\Delta h = 0.1$ cm; $\Delta p = 3.8$ torr; $p_s = 760$ torr

d (cm)	$V_A - V_C$ (ml) for $E = 0.1°C$	$h_{min} - h_{max}$ (cm) 10	30	50
		for v_C (ml)		
0.05	0.535	0.392	1.178	1.96
0.10	2.14	1.57	4.71	7.85
0.20	8.55	6.28	18.85	31.4
0.30	19.3	14.2	42.6	71.0
0.50	53.5	39.2	117.8	196.0

PROCEDURE

For the most general use of gasometric determinations, it is suitable to use two apparatuses: one for determination of active hydrogen and the other for all other determinations. For an apparatus with a manometer working space of about 2 ml and $h_{max} - h_{min} = 50.0$ cm, the procedure is as follows:

0.05 to 0.01 mmole (based on the amount of gas evolved or consumed) of the material to be analysed is weighed into boat *S*. Liquids are weighed into very thin-walled capillaries. The required amount of reaction solution ($LiAlH_4$ solution, catalyst, acid, etc.) is introduced into the reaction space *A*. When the thermostat temperature has been adjusted to within ±0.1°C, the boat *S* is attached. If necessary (i. e., in determinations of active hydrogen or in catalytic hydrogenations), air is displaced from the apparatus by passage of the corresponding gas (hydrogen, nitrogen, etc.) at a rate of about 0.1 ml/sec. This takes about 15 minutes. The manometer-liquid level is then adjusted to the appropriate mark, i. e., in gas-consuming determinations, the manometer is filled with gas, and in gas-evolving determinations, the gas is displaced from the manometer. One then observes whether the measuring-liquid level remains constant, thereby testing the temperature constancy and gas-tightness of the apparatus. There must be no change for 15 minutes. In hydrogenations, the catalyst is saturated with hydrogen during this phase, until the liquid level again remains unchanged for the interval stated. The saturation times differ, depending on the nature of the catalyst and solvent. After saturation of the catalyst, the measuring capillary must again be filled with gas. By rotating the boat about the ground joint, the sample is introduced into the reaction solution. In the analysis of a liquid, the magnetic stirrer is accelerated at this point until the capillary breaks. The liquid level in the manometer is again observed, and is read after it has remained constant for 15 minutes. The result is calculated from equation (1). Immediately after the determination, the reaction space is cleaned by means of a suction pipette.

Equation (1) is simplified for determinations in which the solubility of the gas in the reaction solution is negligibly low. For 5 ml of reaction solution, this criterion is expressed by a gas solubility of $7.5.10^{-4}$ mmoles/ml (at 25°C and 760 torr).

The results of some determinations are given in Table 2.

TABLE 2

Some Examples of Gasometric Determinations

Material	mg Sample	Reaction medium	Gas Measd.	mg Found	% Error
Mg	1.532	5% HCl	H_2 *	1.528	-0.26
	1.044			1.039	-0.48
	0.573			0.570	-0.50
	1.456	20% CH_3COOH		1.451	-0.34
	0.988			0.985	-0.30
	0.495			0.493	-0.40
$BaCO_3$	15.632	5% HCl	CO_2 **	15.585	0.0
	11.218			11.188	0.27
	8.327			8.306	0.31
$(CH_3COO)_2UO_2$	30.235	50% CH_3COOH	H_2 *	30.133	-0.33
	25.614	+ Pt catalyst		25.510	-0.43
	20.387			20.301	-0.42
$UO_3 \cdot H_2O$	25.424	85% H_3PO_4	H_2 *	25.323	-0.40
	20.621	+ Pt catalyst		20.510	-0.43
	18.635			18.602	-0.18
o-$HOC_6H_4NO_2$	3.256	98% CH_3COOH	H_2 *	3.246	-0.31
	2.834	+ Pt catalyst		2.842	+0.28
	2.042			2.033	-0.44
C_6H_5COOH	8.421	$LiAlH_4$ in	H_2 *	8.421	±0.00
	6.077	dibutyl		6.069	-0.21
	4.331	ether		4.320	-0.25

* Calculated without consideration of gas solubility

** Calculated with consideration of gas solubility

CONCLUSION

These results confirm the justification of the theoretically derived construction parameters and also the broad applicability of the apparatus. The exact values of the parameters for equation (1) are determined either gravimetrically or empirically by a series of determinations of a test substance (e. g., Mg). The error of the determinations does not exceed ±0.5% absolute.

Useful ratios of the construction parameters are most easily obtained by a suitable choice of the volume of reaction solution.

SUMMARY

A universally applicable apparatus for gasometric determinations is proposed, based on mathematically derived construction parameters. Several basic requirements, such as minimum dead space, minimum number of ground joints, good temperature constancy, and independence of the operating procedure of barometric pressure, are satisfied to an outstanding extent by the apparatus specified. Its universal applicability was tested by a series of different determinations. The absolute error does not exceed ±0.5%.

LITERATURE CITED

[1] R. Kuhn and E. F. Möller, *Angew Chem.* **47**, 145 (1934).

[2] W. Lüttgens and E. Negelein, *Biochem. Z.* **269**, 177 (1934).

[3] B. Buděšínský, *Collection Czech. Chem. Commun.* **24**, 2948 (1959).
[4] K. Hozumi, *Yakugaku Zasshi* **80**, 410 (1960).
[5] V. Pechanec and J. Horáček, *Collection Czech. Chem. Commun.* **26**, 1805 (1961).

23

*Microanalytical Hydrogenation**

B. G. Belen'kiĭ and I. S. Baraboshina

The Warburg manometric method [1,2] permits the determination of the quantity of gas formed or consumed in a system at constant temperature and volume, but with changing gas pressure. The Warburg manometric method in gas microanalysis ensures high accuracy of the determination as compared with the volumetric method (measurement of volume change at constant pressure), since small volume changes of the gas in the system are easier to observe by a manometric method than by direct measurement of volume. Thus, Schwabe [3] points out that the use of the Warburg method permits measurement of changes in gas volume with an accuracy of up to ±0.003 cc.

There are directions in the literature for the use of a manometric method of microanalytical hydrogenation. Kuhn and Moller [4] developed a differential manometric method of microhydrogenation, based on measuring the pressure difference between two vessels, in which the sample and a standard compound are hydrogenated under identical conditions. This method, providing an accuracy of measurement of ±0.5% for a 40-hour hydrogenation of a 5-mg sample,

**Zhur. Anal. Khim.* 16, 337-42 (1961)

is characterized by considerable complexity and difficulty of execution. Umbreit, et al. ([1], p. 141) developed a simple manometric method, using a standard Warburg apparatus, for the analytical hydrogenation of pyridoxal with an accuracy of ± 4%.

The object of our investigation was the development of a simple manometric method of microanalytical hydrogenation, using a Warburg apparatus. We succeeded in producing special hydrogenation cells and a method of catalytic heterogeneous hydrogenation providing a possibility for microanalytical determination and a kinetic study of this process.

EXPERIMENTAL PART AND DISCUSSION OF RESULTS

a) Preparation and Purification of Hydrogen

The hydrogen required for the hydrogenation was prepared electrolytically in a Bruns electrolyzer with nickel electrodes [5]. The electrolyte is a solution of 200.0 g KOH and 50.0 g $Ba(OH)_2$ in 500 ml water. The hydrogen formed, after preliminary drying over $CaCl_2$, was freed of traces of oxygen in a tube containing palladized asbestos at 300-350°, and of traces of H_2S in a wash bottle containing a 20% NaOH solution saturated with PbO. After separating the gas from the spray of NaOH solution in a safety trap and tube with glass wool, it was finally dried in a $CaCl_2$ tube.

b) Adaptation of the Warburg Apparatus for Hydrogenation

In our experiments we used a Warburg apparatus made by the Labor company (Hungary), designed for cell respiration studies. This apparatus consists of a water thermostat with turbine stirrer and mercury-toluene thermoregulator, ensuring a bath temperature constant to within ± 0.05° over the range 20-80°. On the side walls of the thermostat there are attached shaking panels, with a maximum of 260 oscillations per minute. The amplitude and speed of oscillation are easily controlled. Warburg manometers are attached to the panels by special clamps, and vessels dipping into the water of the thermostat are connected to the manometers by means of ground joints.

In the course of the investigations we found that, despite the directions of Umbreit ([1], p. 141) for using a standard system of Warburg respirometers for microanalytical hydrogenation, these have the following disadvantages preventing their use: 1) because of the small depth of the retorts [Tr. note: the conventional pistol-shaped container], the material being hydrogenated is prematurely spattered into the working space of the vessel during saturation of the catalyst with hydrogen; 2) during the oscillation, the catalyst collects in a pile at the bottom of the vessel and thus is removed from the reaction; 3) lubricant from the ground joint gets into the hydrogenation vessel; 4) it is extremely difficult to obtain complete replacement of air by hydrogen even on prolonged passage of hydrogen through the manometer into the Warburg vessel; 5) it is impossible to control the degree of saturation of the catalyst by hydrogen during the experiment.

Taking these disadvantages into account, we modified the standard Warburg vessel, and adapted it for hydrogenation (Fig. 1). The working space in the vessel used by us is a horizontal hemicylinder with a flat bottom **1**, whose axis is in the direction of oscillation. To this is joined a vertical tube **2**, ending in a

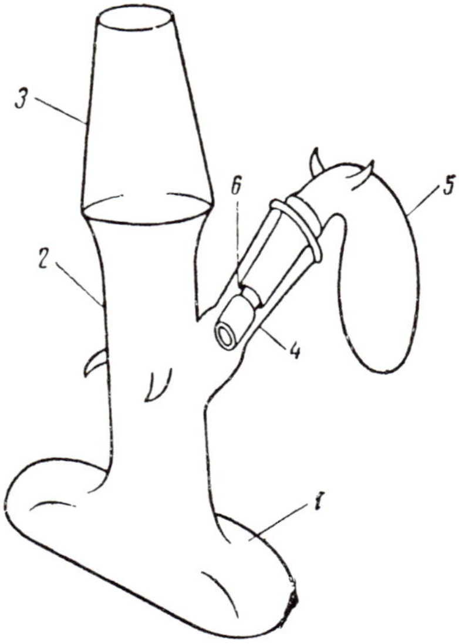

Fig. 1 Hydrogenation vessel. 1: Horizontal part; 2: vertical tube; 3: ground joint; 4: side tube with ground joint; 5: rotatable retort; 6: groove.

male ground joint **3**, which prevents lubricant from entering the vessel. To the vertical part of the vessel there is sealed, at a 60° angle, a tube **4**, to which a rotatable deep retort **5** is attached by means of a ground joint. On the ground zone of the retort is a groove **6**, which protects against lubricant entering the hydrogenation vessel.

The ground joints were lubricated with vacuum grease, prepared by melting vaseline, paraffin, and natural rubber in the ratio 5:2:5. Some changes were also introduced into the construction of the manometer (Fig. 2). To the rubber tube **2** there was attached a two-way stopcock **3**, connected by a rubber tube 4 to a glass bulb **5**. This construction permitted us to evacuate the manometric liquid (mercury) from the manometer and, after turning stopcock **3**, to pump out the gas from the system with a water-jet pump and then to fill it with hydrogen through the three-way stopcock **6**. By alternately evacuating the system and filling it with hydrogen five times, it was possible to remove the air completely and to obtain a hydrogen atmosphere. Stopcock **3** was then closed, the bulb and rubber tube were disconnected, and the apparatus was placed in the starting position. By turning the screw clamp, with the Warburg vessel attached to the manometer and the three-way stopcock open, the level of the manometric liquid in both arms of the manometer was brought to the zero point (the 250-mm mark). The three-way stopcock was then closed, the experiment was run, and the pressure change in the system measured by the change in the level of the manometric liquid in the left arm after bringing the system to constant volume by adjusting the level of the manometric liquid to the zero point in the right arm. The manometer readings were corrected by a so-called thermobarometer—an ordinary manometer with a Warburg vessel filled with the solvent used in the reaction and under the same conditions as the working manometer. This permitted us to take into account the effect of temperature changes in the bath and surrounding air, and also of atmospheric pressure, on the pressure change of the gas in the system during the experiment.

c) Measurement of the Volume of the Hydrogenation Vessel.

The volume of the hydrogenation vessel and the manometer up to the zero point is determined by the method described by Umbreit ([1], p. 85), by weighing the mercury required to fill this space.

d) Calculation of the Amount of Hydrogen Absorbed.

The hydrogen uptake is calculated by the formula (1):

$$x = \left[\frac{(V - V_1)\frac{273}{T} + V_1 \alpha}{P_0} \right] h = Kh \qquad (1)$$

where V is the volume of the hydrogenation vessel and the working part of the manometer (ml); V_1 is the volume of liquid in the vessel (ml); α is the solubility of hydrogen in the liquid at normal pressure (mg/ml); T is the temperature of the experiment in absolute degrees; P_0 is the normal pressure (mm of manometric liquid); h is the pressure drop in the system during the experiment (mm of manometric liquid); x is the hydrogen uptake (ml); K is the flask constant. Since $P_0 = 760 \cdot 13.60/d$, where 760 is the normal pressure in mm Hg, 13.60 is the specific gravity of mercury, and d is the specific gravity of the manometric liquid, we can write, from equation (1):

$$K = \frac{x}{h} = \left[\frac{(V - V_1)\frac{273}{T} + V_1 \alpha}{760 \cdot 13.60} \right] d \qquad (2)$$

and, when mercury is used as the manometric liquid,

$$K = \frac{(V - V_1)\frac{273}{T} + V_1 \alpha}{760} \qquad (3)$$

Since all factors entering into equation (2) are constant

for a given experiment, the determination of the gas uptake reduces to multiplying the flask constant K by the change in the level of the manometric liquid h.

From formula (2) it also follows that the value of K characterizes the sensitivity of the hydrogenation apparatus. To increase the sensitivity, the flask constant K must be reduced. Of particular significance is a decrease in the specific gravity of the manometric liquid, since a change in d from 13.60 (mercury) to 0.80 (96% ethanol) increases the sensitivity of the method by a factor of 17 without changing the construction and geometric dimensions of the hydrogenation cell.

e) Hydrogenation of Maleic Acid over Palladium Oxide.

The efficiency of the apparatus developed was tested on the hydrogenation of triply recrystallized (from doubly distilled water) maleic acid. Palladium oxide was prepared by the method of Adams. [6] A suspension of PdO (0.7-1.6 mg in 1 ml doubly distilled water) was placed in the working part of the hydrogenation vessel (see Fig. 1); a solution of 11.8-12.6 mg maleic acid in 1.2 ml doubly distilled water was placed in the side retort. The system was filled with hydrogen and the catalyst was saturated with hydrogen at 100-120 oscillations per minute. After the hydrogen absorption by the PdO was complete, the mercury level in both arms of the manometer was adjusted to the zero point, after which the retort with the solution of maleic acid was inverted, and hydrogenation was started with a maximum shaking rate of 240-260 oscillations per minute, controlling the course of the process by decreasing the hydrogen pressure in the system. The termination of the hydrogenation is

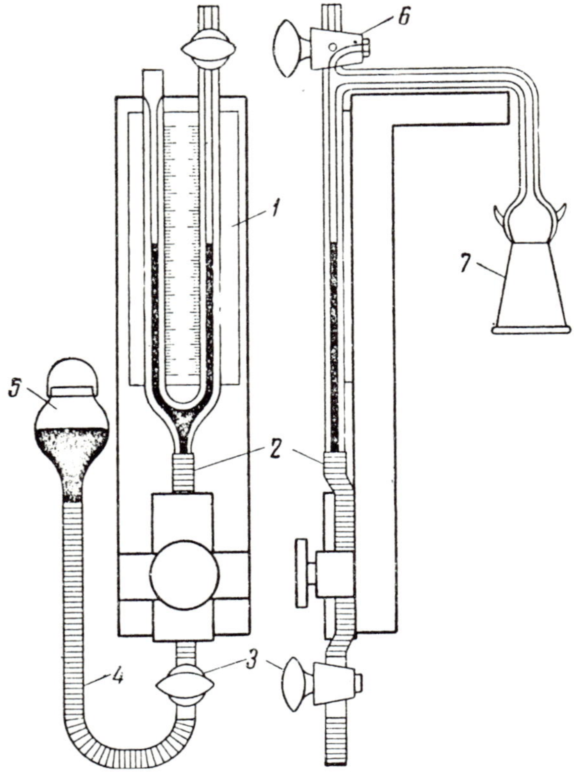

Fig. 2 Modified manometer. 1: U-shaped manometric tube; 2: rubber tube; 3: two-way stopcock; 4: rubber tube; 5: glass bulb; 6: three-way stopcock; 7: branch with ground joint for attachment to the hydrogenation vessel.

judged from the cessation of hydrogen absorption. The experiments were carried out at 40 ± 0.05°.

Results of one of the experiments are represented in Fig. 3, which shows the kinetic curves for saturation of the catalyst with hydrogen and hydrogenation of maleic acid. The measurements show that an increase in the amount of PdO favors an increase in the reaction rate. Thus, a change in the weight ratio of PdO to maleic acid from 0.05 to 0.13 leads to a decrease in the hydrogenation time from 165 to 60 minutes.

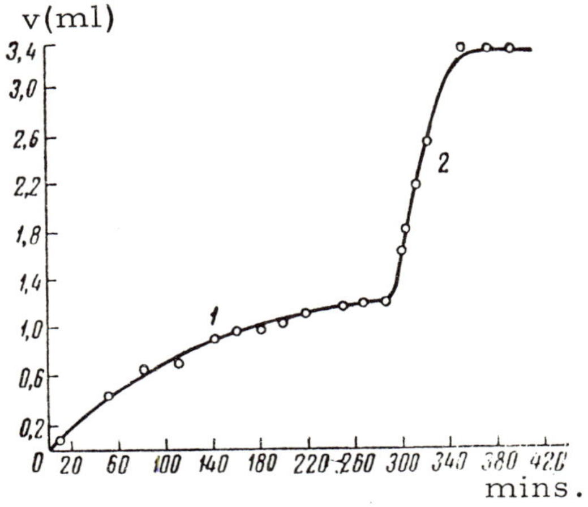

Fig. 3 Hydrogenation of maleic acid over PdO. 1: Saturation of catalyst with hydrogen; 2: hydrogenation of maleic acid.

In the experiments described, there was good reproducibility of the results, which differed among themselves by no more than 0.4%, with a deviation of 0.4-0.8% from the theoretical value of the hydrogenation equivalent, for a hydrogen uptake of 2.21-2.26 ml in a vessel of volume 25.34 ml.

f) Kinetics of the Hydrogenation of Maleic Acid over a Cu-Pd Catalyst.

As is apparent from Fig. 3, the kinetic curve of

the hydrogenation of maleic acid over PdO is first-order. This may be a reflection of the true character of the catalytic process, and also a result of diffusion complications arising at the catalyst surface. The latter assumption is supported by literature data [7, 8, 9] indicating a zero order for the hydrogenation of maleic acid over palladium contacts. To check the possibility of using the proposed method for the study of hydrogenation kinetics, we attempted to eliminate the diffusion complications. To do this, we increased the number of oscillations from 270 to 400 per minute (introducing the necessary changes in the Warburg apparatus). However, this led to spattering of the catalyst into the vertical part of the hydrogenation vessel and to its removal from the sphere of reaction. We therefore decided to fix the catalyst in the working part of the vessel. The catalyst was used in the form of platinum black, applied on platinum sheet or gauze (area 1.7 sq. cm), sealed into the horizontal part of the hydrogenation vessel. (Fig. 4) Instead of PdO, we used a Pd-Cu alloy containing 30 atom per cent Cu, as recommended by Karpova and Tverdovskiĭ. [9] This catalyst has a high activity for the hydrogenation of ethylenic and acetylenic bonds, and is more stable to poisoning than pure Pd. The Pd-Cu alloy was prepared by electrodeposition from a bath consisting of a mixture of the following solutions: I. 0.200 g $PdCl_2 \cdot 2H_2O$ and 0.250 g $NaNO_2$ in 10 ml H_2O; II. 0.14 g $CuSO_4 \cdot 5H_2O$ and 0.360 g $(NH_4)_2SO_4$ in 10 ml H_2O; III. 0.14 N H_2SO_4, in a volume ratio I:II:III = 5.9:4.1:30. The deposition time varied from 3 to 10 minutes at a current density of 70 milliamps/sq. cm. The precipitate of Cu-Pd black was washed with water and subjected to cathodic purification to remove adsorbed ions, at a current density of 70 milliamps/sq. cm, for 30 minutes in 0.1 N HCl or 0.05 N H_2SO_4. The hydrogenation of maleic acid was carried out at 20 and 40°

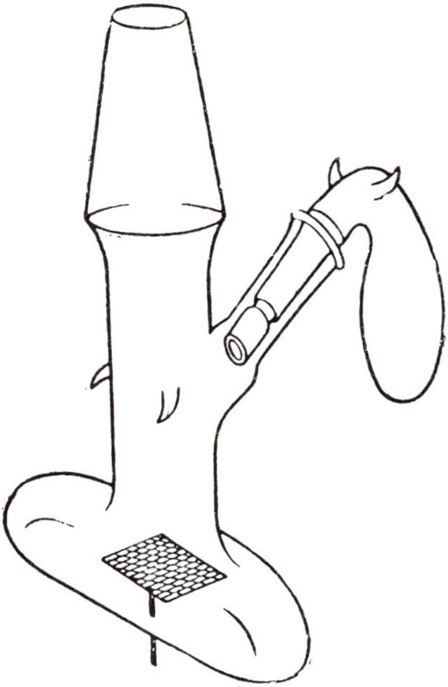

Fig. 4 Hydrogenation vessel with a fixed catalyst.

(± 0.05°) by the method described above. The kinetic curve of hydrogen absorption by the Pd-Cu catalyst is analogous to the curve for PdO in Fig. 3. The amount of hydrogen required to saturate the catalyst is almost five times that required for the hydrogenation of maleic acid. This indicates the need for careful control of the saturation of the catalyst with hydrogen in microanalytical hydrogenation. As in the case of PdO, it was noted that the larger the amount of hydrogen used to saturate the catalyst (i. e., the larger the amount of catalyst used in the experiment), the faster is the hydrogenation. The kinetic curves for the hydrogenation of maleic acid are shown in Fig. 5. In these experiments we also achieved good reproducibility of the results (± 0.1%, with a deviation of + 0. 4%

from the theoretical value of the hydrogenation equivalent, for a hydrogen uptake of 2.05-2.14 ml in a vessel of volume 27.7 ml).

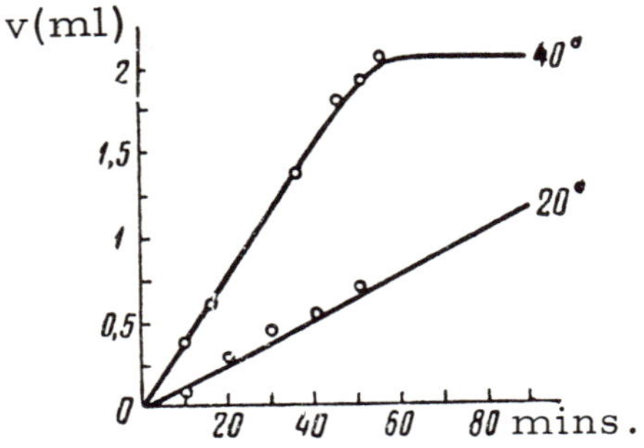

Fig. 5 Hydrogenation of maleic acid over a Pd-Cu catalyst.

As is apparent from Fig. 5, the kinetic curves for the hydrogenation of maleic acid over a Pd-Cu catalyst are of the characteristic shape for a zero-order reaction (rate constant independent of concentration). The reaction rate constant was calculated from the formula $K = \Delta x/\Delta t$ and had the following values, at 20 and 40° respectively: K_1 = 0.014 ml/min and K_2 = 0.039 ml/min. The results obtained permitted calculation of the activation energy:

$$E_A = \frac{4.575\ \Delta \ln K}{\Delta 1/T} = 9,250 \text{ cals/mole}$$

which falls within the limits of the values of 1,000 to 10,000 cals/mole obtained by Kobozev[10] for the hydrogenation of ethylene derivatives. This indicates the possibility of using the microanalytical apparatus developed by us to study the kinetics of the hydrogenation process.

SUMMARY

An apparatus for microanalytical hydrogenation (hydrogenation vessels and manometers) was developed, permitting the quantitative hydrogenation of a compound with hydrogenation equivalent ≅ 100 with an accuracy of ± 0.5%, using a 10-mg sample and mercury as the manometer liquid.

Using maleic acid as an example, it was shown to be possible to use the modified apparatus for studying the kinetics of catalytic hydrogenation.

LITERATURE CITED

[1] W. W. Umbreit, R. H. Burris, and J. F. Stauffer: "Manometric Methods for the Study of Tissue Metabolism" [Russian Edition], Moscow (1951).

[2] N. P. Meshkova and S. Ya. Severin: "Laboratory Manual on the Biochemistry of Animals", Moscow (1950), p. 260.

[3] K. Schwabe, *Z. Electrochem.* **61**, 744 (1957).

[4] R. Kuhn and E. F. Möller: "Analytische Methoden. Methoden der organischen Chemie. II", Stuttgart (1953), p. 296.

[5] B. P. Bruns: "Modern Laboratory Technique", Collection of Papers, Moscow (1935), No. 1, p. 14.

[6] C. Weygand: "Experimental Methods in Organic Chemistry" [Russian Edition], Moscow: 1950, Vol. II, p. 22.

[7] Z. Csuros, et al., *Acta. Chim. Hung.* 1, 22 (1951).

[8] Z. Csuros, et al., *ibid.* 1, 168 (1951).

[9] R. A. Karpova and I. P. Tverdovskiĭ, *Sbornik Trudov. Gos. Inst. Priklad. Khim. po Fiz. Khim.* **No. 42** , Leningrad (1959), p. 212.
[10] N. I. Kobozev, *Usp. Khim.* **25**, 545 (1956).

24

The Use of the Warburg Manometric Method for the Microdetermination of Active Hydrogen and for Catalytic Hydrogenation*

B. G. Belen'kiĭ and L. O. Bol'shakova

Despite the existence of good micro methods of functional analysis, particularly for the determination of active hydrogen [1,2], it is expedient to have methods permitting, for example, the determination of active hydrogen and unsaturation in the same apparatus. A number of such microanalytical methods [3,4] are described in the literature, their use being particularly convenient in small laboratories. Almost all of these methods are volumetric. Just recently, papers have been published by Horáček and Pechanec [5,6], who have proposed a manometric method permitting determination of active hydrogen and hydrogenation in the same cell. The manometric method is more suitable for microdetermination, since it is characterized by great sensitivity; with it, volume changes of 0.001 ml can easily be observed. The micro apparatus for the manometric method is considerably simpler; here only the accurate determination of the volume of the gas space is necessary, whereas, in the volumetric method, a corresponding sensitivity requires microgasometers with 0.005-0.002 ml graduations. Moreover, a simple change in the

**Zhur. Anal. Khim.* 18, 873-79 (1963)

specific gravity of the manometric liquid permits an increase in the sensitivity of the method without changing the apparatus.

Like Horáček and Pechanec, we have used a Warburg apparatus for the manometric determination of active hydrogen and for catalytic hydrogenation. This apparatus is used relatively little in organic analysis. [5-7] However, the equipment is readily available, since it is currently mass-produced in the "Fizpribor" factory in Frunze under the name "AG-1 apparatus for carrying out gas microanalyses".

In the determination of active hydrogen, we have tested several modifications of the method. At first, following the design of the manometer vessel with two side arms which is generally used for these purposes, we have used a manometer cell of the same type. We place the sample to be determined in the main space of the vessel, and the reagent in the rotatable "retort" side arm. As reagent we have tried both a Grignard reagent and $LiAlH_4$. In the former case we have used as solvents for the sample anisole, pyridine, and their mixtures, which have a considerably greater solvent action than pure anisole. [8-10] The best results were obtained with the 1:5 pyridine-anisole mixture recommended by Soltys. [10] The experiments were carried out at 25 and 40°C. The lower temperature is more convenient, since the solvent then has a lower vapor pressure and the equilibrium state is therefore established more rapidly.

The determination with $LiAlH_4$ was carried out initially using dipropyl ether as recommended in the literature. [11] However, this solvent is inconvenient, since it has a poor solvent action and high vapor pressure, which delays the establishment of equilibrium. This also holds true for tetrahydrofuran. In this respect, 4-ethylmorpholine [12] is incomparably better, and better yet are 4-pentylmorpholine and 4-butylmorpholine, which were first used by us. These solvents have a low vapor pressure and good solvent action. The reaction was carried out in a medium of dry, purified nitrogen when using the Grignard reagent, and in dry purified hydrogen in the case of $LiAlH_4$. Using the method described, we have determined hydrogen in 3-5-mg samples with an accuracy of ± 1-3%. Better results were obtained with $LiAlH_4$.

The experiments as initially modified however, had a substantial drawback, in that they included carrying out a blank determination, which, besides increasing the time of the experiment, made it necessary to stabilize the experimental conditions: temperature, stirring rate, time of readings, etc. Attempts to eliminate the blank in the case of $LiAlH_4$, as recommended in the literature, led to unsatisfactory results, since, at the moment of pouring the reagent solution from the "retort" side arm to the main space, a pressure increase occurred, apparently on account of reaction with the moisture adsorbed on the walls and through an increase in the rate of evaporation of the solvent as a result of the increase in its surface area.

The latter disadvantage also occurs in the method of Horáček and Pechanec [5], which obliges them to introduce a correction, determined in a blank test. It is obvious that better results can be obtained by not using a vessel with two side arms. However, the use of the cell described in the literature [13], which is a pear-shaped flask with a rotatable spoon in the ground-glass stopper, was unsuccessful. In this case, some samples such as benzoic acid or diphenylamine, when placed on the spoon, dissolved even before the start of the reaction, evidently because of absorption of solvent, and unsatisfactory results were obtained. We were able to avoid these difficulties by using a cell of a new design.

The cell developed by us (Fig. 1) consists of a vessel **1** (capacity about 10 ml) with a ground joint **2** for connection to a manometer. The vessel has a side arm **3**, provided with a stopcock **4**. At the bottom of the vessel is a rotatable ground stopper **5**, having a notch **6** in its end. The circular grooves **7** on the stopper are a protection against entry of the lubricant into the vessel. The reagent with the $LiAlH_4$ is poured into the bottom of the vessel, and the tip of the capillary or sealed ampoule (in the case of volatile liquids) containing the weighed sample is placed in the notch in the stopper.

Upon attainment of equilibrium, which required 15-20 minutes, the capillary or ampoule was broken by turning the stopper, and the active hydrogen content was determined from the pressure increase in the manometer cell. This method has permitted us to eliminate the blank and to limit ourselves to a 10-minute passage of hydrogen through the system. We have determined active hydrogen in 1.5-3-mg samples with an accuracy of ± 1-3%. The whole determination in this case required 30-40 minutes.

The use of the Warburg method for analytical hydrogenation [6, 7, 14] is more widely known than its use for the determination of active hydrogen [5]. Analytical hydrogenation in a Warburg apparatus is particularly expedient for studying the kinetics of the process. Thus, it was shown that it it much easier to reach the kinetic region with shaking than by stirring with a rotary stirrer.

The manometer cell described by us (Fig. 1) was also very convenient for microanalytical hydrogenation. As catalyst we have used palladium on charcoal [15] and Raney nickel [16]. Hydrogenation was

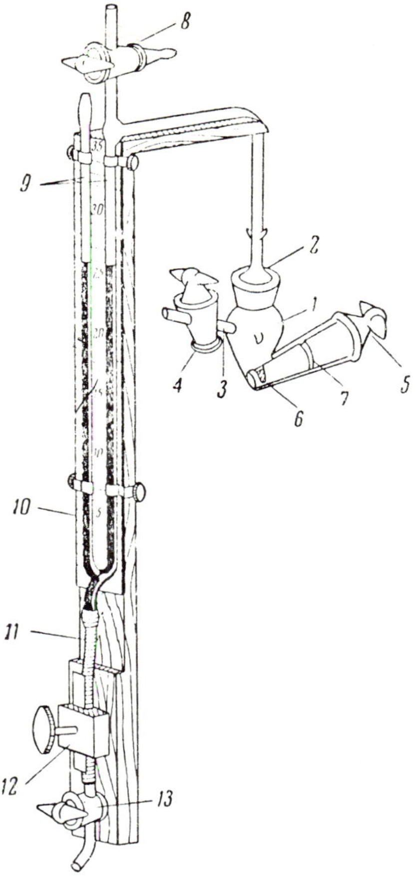

Fig. 1 Universal manometer cell. 1: Manometer vessel; 2: ground joint for connection to manometer; 3: side arm; 4: stopcock; 5: ground rotatable stopper; 6: notch for placement of capillary or ampoule; 7: circular groove; 8: stopcock for connection to atmosphere; 9: U-shaped manometer; 10: millimeter scale; 11: rubber tube; 12: screw clamp for adjusting the mercury levels in the manometer; 13: stopcock for filling the manometer with mercury.

carried out in alcohol and dimethylformamide solutions, with 1.5-3-mg samples. The accuracy of the determination is ± 1-2%.

EXPERIMENTAL PART

PURIFICATION OF HYDROGEN

Electrolytic hydrogen from a cylinder is collected in a gasometer, passed through two bottles with 3% $KMnO_4$ in 10% KOH, and dried in a $CaCl_2$ tube; the hydrogen is then passed through a tube with palladized asbestos heated to 350°C, then through a washer made of a multiply-bent tube filled with a solution of lead oxide in 20% NaOH, to remove H_2S and CO_2, and, finally, the hydrogen is dried by passing it through a $CaCl_2$ tube, Anhydrone column, and U-shaped P_2O_5 tube.

PREPARATION AND PURIFICATION OF 4-PENTYLMORPHOLINE AND 4-BUTYLMORPHOLINE.

Equimolar amounts of freshly distilled bis(2-chloroethyl) ether and pentylamine or butylamine (0.5 mole) in 250 ml of 40% aqueous KOH solution are heated under reflux on a glycerol bath for 11 hours, gradually raising the temperature from 100 to 140°C in order to ensure sufficiently intensive boiling of the solution. The upper layer is separated; the lower layer, containing the crystalline precipitate, is extracted with diethyl ether, and the extract is added to the upper layer. The ether is distilled off. The pentyl- or butylmorpholine is extracted several times with 3.5 N HCl.

The HCl solution is made alkaline with 40% KOH,

and the pentyl- or butylmorpholine is extracted from the alkaline solution with diethyl ether. It is dried over solid KOH, the ether is distilled off, and the residue is vacuum-distilled. For pentylmorpholine, the fraction boiling at 57-58°C (5 mm Hg) is used; for butylmorpholine, the fraction boiling at 59-60°C (10 mm Hg). This fraction is dried further over ignited barium oxide and redistilled, collecting the fraction boiling at the same temperature. The yield is 35-40%. For 4-pentylmorpholine, $n_D^{20} = 1.4469$; $d_4^{20} = 0.8908$; b. p. = 212-213°C (758 mm); $R_D = 47.145$; R_D (calc.) = 47.145 [17]. For 4-butylmorpholine, $n_D^{20} = 1.4463$; $d_4^{20} = 0.8967$; b. p. = 167-168.5°C (762 mm Hg); $R_D = 42.48$; R_D (calc.) = 42.527 [17].

PURIFICATION OF ETHANOL

96% ethanol is refluxed over ignited $CuSO_4$, filtered, and distilled twice from CaO.

PURIFICATION OF DIMETHYLFORMAMIDE

150 g P_2O_5 is added to 1 liter of dimethylformamide. After standing for 2-3 days, the mixture is distilled 3-4 times at 20-30 mm Hg.

DETERMINATION OF ACTIVE HYDROGEN

A 1.5-3-mg sample is placed into a weighed capillary or ampoule (Fig. 2). The capillary or ampoule is placed in the notch of the stopper. Into the vessel is poured 0.7 ml of a suspension of $LiAlH_4$ in butyl- or pentylmorpholine, prepared 10-12 hours before the experiment, and the walls of the vessel are rinsed with the reagent. The vessel is then connected to a mercury manometer, and a stream of hydrogen is passed through the system for 10 minutes, passing it through the stopcock 8 of the manometer and the side arm 3 of the vessel. The side arm is connected to a seal filled with glycol. The system is then placed in the thermostat of the Warburg apparatus at 25 ± 0.05°C.

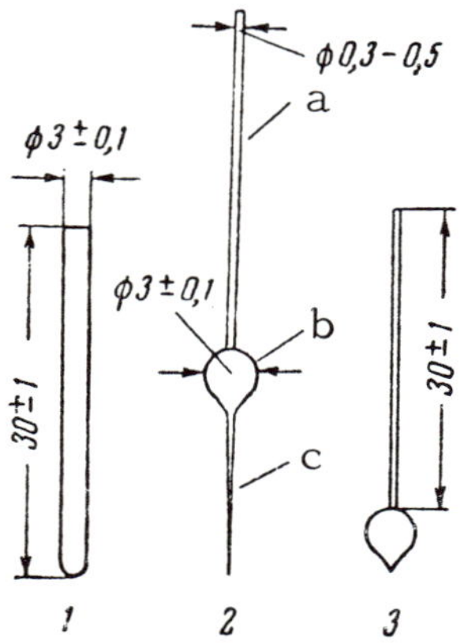

Fig. 2 Capillary and ampoule for samples. 1: Capillary for introducing solid sample; 2: ampoule for introducing liquid; a: glass rod, b: ampoule, c: capillary to collect liquid; 3: sealed ampoule.

After shaking for 5 minutes, stopcock 8 is rapidly opened and the pressure in the system is equalized with the atmospheric pressure. At 100-120 oscillations per minute, the manometer is read every 5 minutes, bringing the mercury level in the right arm to the zero point selected for the manometer.

The termination of the reaction is judged by the attainment of a steady pressure increase (1-1.5 mm in 5 minutes).

The pressure change related to the reaction of the sample with $LiAlH_4$ is calculated from the formula:

$$h = h_t - h_0 - \frac{\Delta h \cdot t}{5} \qquad (1)$$

where h is the change in pressure (mm) due to reaction of the active hydrogen, with a thermobarometric correction [14]; h_t is the pressure (mm) at the end of the experiment; h_0 is the pressure (mm) at the moment of crushing the capillary (start of the reaction); Δh is the mean pressure change (mm) in the system per 5 minutes (before the start of the reaction and after the end of the reaction, due to reaction of $LiAlH_4$ with the solvent and for other reasons); t is the reaction time in minutes.

Calculations

It is known that the reaction of $LiAlH_4$ with one atom of active hydrogen liberates one molecule of hydrogen:

$$R\text{-}H + \tfrac{1}{4}\,LiAlH_4 \rightarrow R\text{-}(LiAl)_{\frac{1}{4}} + H_2{}^{*}$$

The volume of gas liberated is calculated from the formula:

$$x = h\left[\frac{(V-V_1)\,\frac{273}{T}}{760}\right] = h \cdot k,$$

where x = volume change in ml; V = volume of system in ml; V_1 = volume of reagent and glass capillary [= (weight of capillary) /(specific gravity of glass (=2.5))]

---

* $LiAlH_4$ also reacts with a nitro group; in this case, two molecules of hydrogen are liberated:
$2R\text{-}NO_2 + 2\,LiAlH_4 \rightarrow R\text{-}N{=}N\text{-}R + 4H_2 + 2LiAlO_2$

in ml*; T = thermostat temperature; h = pressure change, with the thermobarometric correction;

$$k = \frac{(V - V_1) \cdot \frac{273}{T}}{760} = \text{vessel constant}^{**}$$

Thus, when the experiments are carried out under identical conditions, the volume of hydrogen liberated is determined by multiplying h by k. Since the gas medium used is hydrogen, which saturates the solvent before the start of the reaction of $LiAlH_4$ with the sample, we can neglect the solubility of the gas in the solvents, which enters into the usual formulas for Warburg calculations. [14]

The percentage of active hydrogen is calculated from the formula:

$$\%\text{H} = \frac{0.0449\ x}{m} \cdot 100,$$

---

*When using capillaries of the same width (± 0.1 mm) and length (±1 mm), the volumes of glass in the capillary were equal within 10 %, which gave an overall error of about 0.1% in the determination; it was therefore not necessary to calculate k in each experiment. The volume of the sample was not taken into account, since it was less than one-thousandth of the total volume of the vessel.

**In the reduced formula, the number 760 represents the atmospheric pressure, expressed in mm Hg. When a different manometer liquid is used, the denominator has the form 760·(13.6)/d, where 13.6 = specific gravity of mercury and d = specific gravity of the manometer liquid used. It is evident that, by lowering d, k can be reduced, and the sensitivity of the method thus increased.

where 0.0449 is the weight of 1 ml of hydrogen in mg and x is the reduced volume of the liberated hydrogen in ml. Since k is the same in all experiments, the formula takes on the form:

$$\%\mathrm{H} = \frac{h}{m}\,(4.49k)$$

The results of experiments on the determination of active hydrogen, using $LiAlH_4$, are shown in Table 1.

CATALYTIC HYDROGENATION

A suspension of 30 mg of catalyst in 0.7 ml of solvent is introduced into the reaction vessel, the capillary with the sample (1.5-3 mg) is placed in the notch of the rotatable stopper, the vessel is connected to the manometer, and a stream of hydrogen is passed through for 10 minutes. The vessel with manometer is then placed in a thermostated bath at 30 ±0.05°C. After constant pressure is reached, which requires 15-20 minutes, the capillary is broken by turning the stopper, and hydrogenation is started. The course of the hydrogenation is followed by the decrease of pressure in the manometer cell.

The results of the hydrogenation experiments are shown in Table 2.

CALCULATIONS

The volume of hydrogen absorbed is calculated from formula (2), and the hydrogenation equivalent from the formula:

$$E_{\mathrm{H}} = \frac{m \cdot 22.41}{x} = \frac{m}{h}\left(\frac{22.41}{k}\right) \qquad (3)$$

TABLE 1

Compound	mg Sample	No. of active H-atoms Calc.	No. of active H-atoms Found	Time to reach equilibrium (min.)
Aniline	0.82	2.0	2.0	20
	2.48		1.98	25
Butylamine	1.39	2.0	2.04	10
	0.87		2.05	10
Benzoic acid	1.47	1.0	1.01	5
	2.03		1.0	15
Salicylic acid	1.95	2.0	2.05	5
	1.88		1.98	10
Acetanilide	2.75	1.0	0.99	10
	2.05		1.03	10
Diphenyl-	3.22	1.0	0.97	15
amine	3.27		0.98	10
1-Naphthyl-	1.68	2.0	1.9	30
amine	1.64		1.94	30
Phenoxy-	3.01	2.0	2.1	25
methylpen-	2.61		2.03	40
icillin (acid)				
Chloram-	1.84	5.0*	5.0	35
phenicol	1.65		4.87	50

*The chloramphenicol molecule contains three atoms of active hydrogen and one nitro group.

where m = sample weight in mg, 22.41 = volume of 1 mg-equivalent of hydrogen in ml; x = reduced volume of absorbed hydrogen in ml; h = pressure change during the experiment, with thermobarometric correction, in mm; k = vessel constant.

The kinetic curves of two experiments on the hydrogenation of maleic and sorbic acids are shown in

TABLE 2

Compound	mg Sample	Catalyst	Hydrogenation equivalent Calc.	Found
Maleic acid	2.30 2.21	Pd/C	116	115.6 115.3
Sorbic acid	1.75 2.03	Pd/C	56.06	56.6 57.2
Methacrylic acid dimethyl-hydrazide	2.95 3.30	Pd/C	128.15	127.60 127.90
Nystatin	2.39 2.11	Pd/C	158.0	163.0 160.0
Croton-aldehyde	1.48 2.35	Pd/C	70.1	71.1 71.1
Croton-aldehyde	2.1 2.36	Raney nickel	35.05	35.8 35.6

Figs. 3 and 4. As can be seen from Fig. 3, the kinetic curve of the hydrogenation of maleic acid has the characteristic form for a zero-order reaction, indicating attainment of kinetic experimental conditions.

SUMMARY

A design has been developed for a universal cell permitting determinations of active hydrogen and catalytic hydrogenation to be carried out without a blank on 1.5-3-mg samples with an accuracy of ± 1-3% for active hydrogen and ± 1-2% for hydrogenation.

It is suggested that butylmorpholine and pentylmorpholine be used as solvents for the determination of active hydrogen by means of lithium aluminum hydride.

The proposed method is suitable for the study of hydrogenation kinetics.

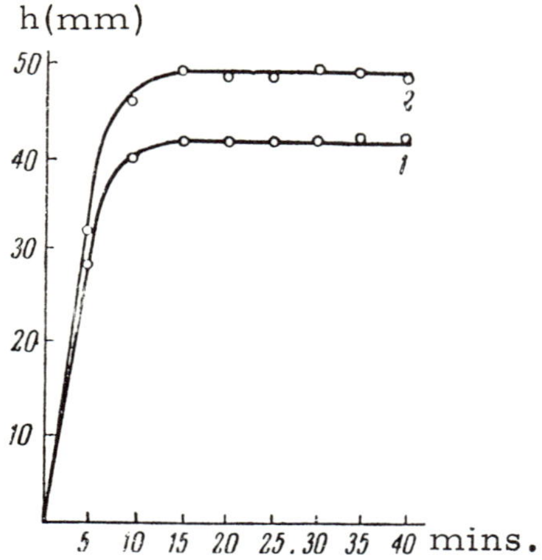

Fig. 3 Hydrogenation of maleic acid on Pd/C.
1: m = 2.21 mg; 2: m = 2.66 mg.

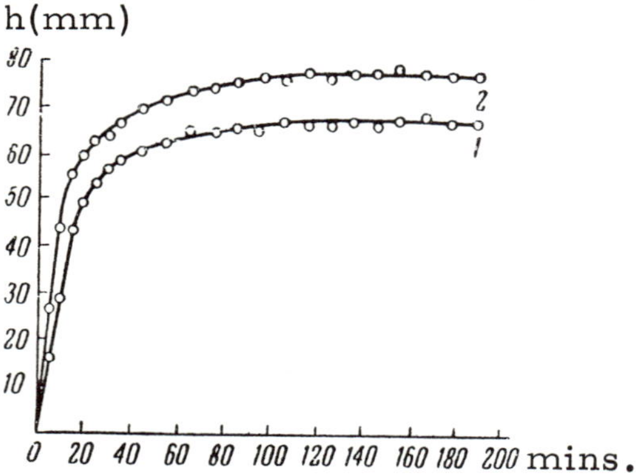

Fig. 4 Hydrogenation of sorbic acid on Pd/C.
1: m = 1.75 mg; 2: m = 2.03 mg.

LITERATURE CITED

[1] A. P. Terent'ev and A. I. Kireeva, *Izv. Akad. Nauk SSSR, Otd. Khim. Nauk* **1951**, 172.

[2] M. N. Chumachenko and L. B. Tverdyukova, *Dokl. Akad. Nauk SSSR* **142**, 612 (1962).

[3] M. R. Chapkekar and T. S. Gore, *Mikrochim. Acta* **1959**, 664.

[4] B. Buděšínský, *Collection Czech. Chem. Commun.* **24**, 2948 (1959).

[5] V. Pechanec and J. Horáček, *ibid.* **26**, 1805 (1961).

[6] J. Horáček and V. Pechanec, *ibid.* **27**, 1500 (1962).

[7] B. G. Belen'kii and I. S. Baraboshina, *Zhur. Anal. Khim.* **16**, 337 (1961).

[8] J. A. Krynitsky, J. E. Johnson, and H. W. Carhart, *J. Am. Chem. Soc.* **70**, 486 (1948).

[9] C. Weygand: "Experimental Methods in Organic Chemistry" [Russian Ed.], part 3, Moscow (1950).

[10] A. Soltys, *Mikrochemie* **20**, 107 (1936).

[11] D. Subba Rao, G. D. Shah, and V. S. Pansare, *Mikrochim. Acta* **1954**, 81.

[12] R. P. Linstead, J. A. Elvidge, and M. Whalley: "Modern Methods of Investigation in Organic Chemistry" [Russian Ed.], Moscow (1959).

[13] H. Lieb and W. Schöniger, *Mikrochemie ver. Mikrochim. Acta* **35**, 400 (1950).

[14] W. W. Umbreit, R. H. Burris, and J. F. Stauffer: "Manometric Methods for the Study of Tissue Metabolism" [Russian Ed.], Moscow (1951).

[15] C. L. Ogg and F. J. Cooper, *Anal. Chem.* **21**, 1400 (1949).

[16] B. N. Dolgov: "Catalysis in Organic Chemistry", Moscow (1949), p. 53.

[17] A. Weissberger (Ed.): "Physical Methods of Organic Chemistry" [Russian Ed.], vol. I, Moscow (1950).

25

*The Use of the Manometric Method for the Microdetermination of Active Hydrogen**

V. Pechanec and J. Horáček

The significant reactivity of the active hydrogen of organic substances with various reagents has permitted development of a number of methods for its quantitative determination; for example, methods based on liberation of the gas after reaction with organometallic reagents [1-8] or $LiAlH_4$ [9-11], volumetric determination [12], acylation reactions [13], or, finally, isotopic analysis by means of deuterium [14] or tritium [15]. The most usual method of determination is based on the reaction of the Grignard reagent or, recently, $LiAlH_4$ with the solution of the substance to be analyzed in an inert solvent, to produce methane or hydrogen, whose quantity is usually measured gasometrically. But methods have also been described in which the liberated gas is burned according to the principle of organic elemental analysis, and the products formed are determined gravimetrically [16] or volumetrically [17].

But, since even the most carefully dried apparatus and the solvent contain traces of moisture, the result in these methods of determination is influenced by the accuracy of the blank determination which, however, in view of the character of its causes, need not

Collection Czechoslov. Chem. Communs.* **26, 1805-13 (1961)

be reproducible in a satisfactory manner. Efforts to eliminate the blank have led to the development of methods in which the substance to be analyzed reacts directly with the reagent solution, without solution in an inert solvent. [9, 18] These determinations are limited to a certain degree by the solubility of the substance being analyzed in methylmagnesium iodide or $LiAlH_4$ solutions. Further difficulties appear when working with volatile liquids, which can be weighed accurately only in glass capillaries. Subsequent crushing of the capillary in the reaction vessel, however, e. g., with glass balls [19], presents difficulties.

All equipment used is mostly derived from the original simple Zerewitinoff apparatus and presents the same difficulties, such as dependence on temperature fluctuations and barometric pressure during the analysis, and the not quite accurate reading of the volume of evolved gas in the burette.

It was therefore our intention to develop a simple micro method for the microdetermination of active hydrogen in solid and liquid organic substances, which has none of the mentioned defects and which also permits reduction of the sample weight to 0.5-2 mg. Most suitable for these requirements is the manometric method of Warburg, in which the quantity of gas set free by reaction of the active hydrogen in the sample with the Grignard reagent or $LiAlH_4$ is calculated from the pressure change at constant volume of the reacting system. Other methods for the determination of active hydrogen, e. g., the acylation method, are used only to a limited extent. Isotopic analysis again requires complicated apparatus.

The Warburg manometric method so far has been used for the determination of active hydrogen in samples of about 1 mg only by Lüttgens and Negelein [20], who calculated the quantity of methane formed from the simultaneous change in pressure and volume of the gas in the reaction vessel, since they still used manometers without containers for the manometric liquid. Their method, however, also requires the carrying out of a blank determination. To eliminate a blank and reduce the solubility of the sample in the reagent to a minimum, a solvent must be used that dissolves not only the great majority of organic substances, but also $LiAlH_4$; the latter was used by us because of its generally higher reactivity with active hydrogen.[9, 21] Of the usable solvents, e. g., diethyl through diamyl ethers, dioxane, tetrahydrofuran, N-ethylmorpholine, and the ethers of ethylene glycol and diethylene glycol [22], tetrahydrofuran proved to be the best. By using a modified reaction vessel, provided with a simple crushing device, we were able to determine active hydrogen also in volatile liquids, which were weighed into the capillary.

EXPERIMENTAL PART AND RESULTS

REAGENTS

The solution of $LiAlH_4$ in tetrahydrofuran was prepared by dissolving 3 g of finely ground $LiAlH_4$ in 100 ml tetrahydrofuran, which had been dried by refluxing for several hours with metallic sodium and distilled. The stock solution was stored together with the rest of the undissolved $LiAlH_4$ in tightly closed ampoules sealed with paraffin. Prepurified nitrogen of electric-lamp grade is dried with anhydrous magnesium perchlorate.

APPARATUS

The apparatus consists of Warburg manometers of the currently used type [23] with a millimeter scale, a rectangular 18-l thermostat bath containing an electric heater connected in circuit with a sensitive mercury contact thermometer and a relay, and a stirrer. The bath temperature is 25°C and is kept within ±0.02° of this value. Mixing of the contents of the manometer vessels is carried out through reciprocal motion with adjustable oscillation frequency.

Manometer Liquid

Brodie solution [23] cannot be used for the microdetermination of active hydrogen, since the evaporating water reacts with the $LiAlH_4$ solution to form hydrogen, and also gradually increases the density of the manometer liquid. The requirements for a manometer liquid, which must be a uniform substance without active hydrogen content, with low vapor pressure, density, and viscosity, are met to a satisfactory extent by silicone oil. In our work we used a methylphenylpolysiloxane fraction with density $d_4^{20} = 0.973$ and viscosity $\eta = 3.637$ centistokes/18.5°, which does not boil in the vacuum of a water-jet pump up to 260°C. For convenience in reading, the manometer liquid is dyed with Sudan yellow or another soluble dye. The containers for the manometer liquid are made from pieces of polyethylene tubing, since those made of rubber corrode with silicone oil.

Manometer Vessels

The Warburg standard manometer vessel used in the determination of active hydrogen by Lüttgens and

Negelein [20] and a whole series of further modifications [23] were not suitable for our purposes, since they contain dead space from which it is difficult to displace air with the dry nitrogen. Moreover, these vessels are relatively low, so that during shaking or during mixing of the substance being analyzed with $LiAlH_4$ solution, the latter can easily come in contact with the lubricant on the ground joints, liberating hydrogen, a phenomenon that is observed on the contact of the majority of lubricants with the reagent. Also, the crushing of the capillaries used to weigh out the samples of liquid being analyzed is difficult in the hitherto described manometer vessels.

We have therefore proposed a new form of manometer vessel, eliminating the deficiencies cited. (Fig. 1) The dead spaces of our vessel are minimal, and the extended necks of the flask and side arm *a* prevent contact of the reagent with the ground-joint lubricant during shaking. The side arm of the manometer vessel is suitably modified for crushing the capillaries. In the bend of the side arm there is a small bulge *b* with 3 or 4 points on the inner wall, produced by puncture. The ground stopper *c* is drawn out into a small sealed-off tube, with two simple blades on the end. The length of the tube is chosen such that the blades are about 1 mm from the points when the stopper is inserted. By turning the stopper, the capillary inserted in the side arm is pushed against one of the points by the blade and is easily crushed. The volume of the manometer vessel proposed by us is about 14 ml. The calibration is carried out in the usual manner with mercury [23]. Lanolin was used to lubricate the ground joints.

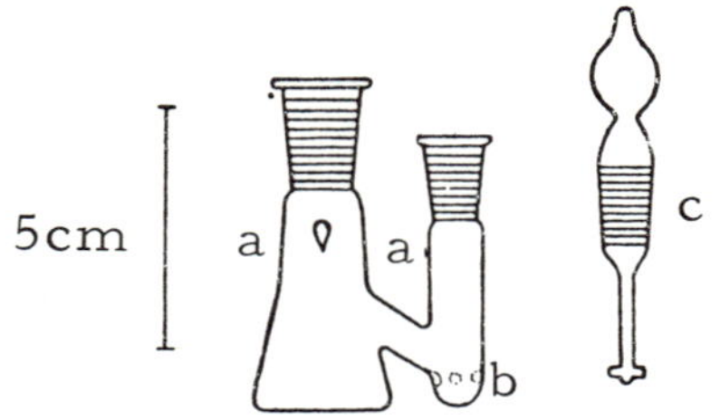

Fig. 1 Manometer vessel. *a*: Extended necks of flask and side arm; *b*: bulge and points; *c*: stopper drawn out to a tube with blades for crushing capillaries.

PROCEDURE

The thoroughly washed and dried manometer vessel, while still warm, is placed on the dry manometer ground joint and is secured with spiral springs.

In order to decrease the dead space of the actual manometer and, at the same time, to prevent squirting out of the manometric liquid during the flushing with dry nitrogen, the side arm, which is connected to the atmosphere, is closed with a stopper, and the manometric liquid is forced into the upper part. The inlet tube for dry nitrogen (15-20 ml/min) is attached to the second arm, and the manometer vessel is flushed for 15 minutes. During this time, the substance to be analyzed is weighed out. Solid samples are weighed into small glass beakers (height and diameter 2 mm), and liquids into small, thin-walled capillaries. After flushing the apparatus with nitrogen for 15 minutes, the weighed sample is placed into the side arm of the manometer vessel and is connected to the nitrogen supply, so that the nitrogen now flows through the vessel in the reverse direction. The vessel is then detached from the ground joint and, in a continuous nitrogen stream, exactly 1 ml of a solution of $LiAlH_4$ in tetrahydrofuran is pipetted into the vessel with a dry pipette in such a manner that the inner wall of the main space is rinsed by it. In the analysis of substances weighed into capillaries, the inner wall of the side arm of the manometer vessel and the capillary itself are also rinsed, to remove moisture adsorbed on the walls (Fig. 2). The ground zone of the manometer joint is then lubricated with lanolin and, in a continuous nitrogen stream, the vessel is placed on it firmly and fastened with springs. The manometer stopcock is now closed, the nitrogen stream is turned off, and the side arm is rapidly closed with the lubricated ground stopper. The stopper is removed from the side arm, the manometer is placed in the thermostated bath, and the shaking apparatus is started. If the initial pressure increase is too great, the excess pressure in the system is reduced by opening the manometer stopcock.

After 60-90 minutes of continuous equilibration, during which the pressure increase, as explained below, is constant and amounts to a maximum of 2-3 mm of barometer liquid column in a half hour, the liquid level in the open side arm is adjusted to the lower limit of the scale, the height is read off, and the actual reaction is carried out. The manometer is taken out of the bath and, by careful tilting, the $LiAlH_4$ solution from the main space of the vessel is transferred to the weighed sample in the side arm (the capillary is crushed by turning the stopper in the side arm), and the manometer is returned to the bath. Since the reaction between $LiAlH_4$ and the active hydrogen of most organic substances is almost instantaneous, the pressure change can already be read off after 5 minutes.

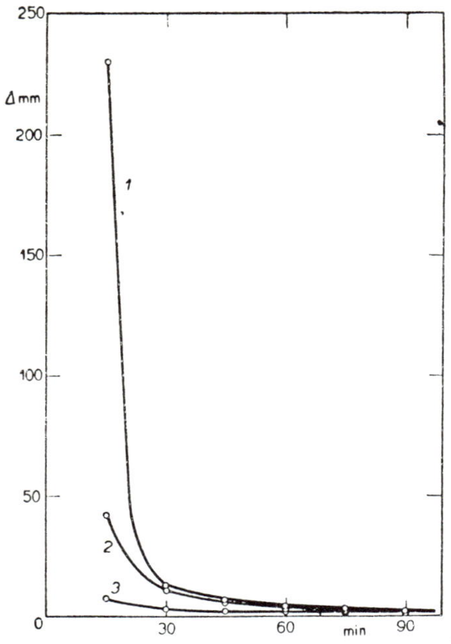

Fig. 2 Dependence of difference in pressure increase on time for various manners of pipetting the $LiAlH_4$ solution into the manometer vessel. 1: $LiAlH_4$ solution is pipetted into the bottom of the manometer vessel; 2: only the inner walls of the main space of the manometer vessel were rinsed with the LiAlH solution; 3: the whole interior of the manometer vessel was rinsed with the $LiAlH_4$ solution.

Only with amides of carboxylic acids and with primary amines is the reaction slower; and the final increase can therefore be read off only after shaking for 10-15 minutes. The active hydrogen content is calculated from the recorded pressure increase.

Calculation

For the calculation, it is necessary to know the relationship between the pressure, expressed in mm of manometer liquid column, and the quantity corresponding to the amount of hydrogen. This ratio is called the vessel constant k. But, since it is more advantageous to express the measured pressure change directly in mg of hydrogen, instead of microliters, half the value of the specific gravity of hydrogen under standard conditions is incorporated into the vessel constant, since only half the amount of hydrogen liberated corresponds to the active hydrogen in the sample being analyzed. If this expanded constant K is multiplied by the pressure change in the manometer in mm of manometer liquid column h, the active hydrogen content of the sample is obtained directly in mg:

$$\%H_{Active} = \frac{hK \cdot 100}{\text{mg sample}}$$

A few results are shown in Table 1. As can be seen from the results shown, the greatest deviations are never more than $\pm 0.02\%$. Ten determinations a day can be carried out by means of the method described.

OTHER CAUSES OF PRESSURE CHANGES IN THE APPARATUS

In the analysis of solid substances, where the walls of the side arm cannot be rinsed with the pipetted $LiAlH_4$ solution, a certain moisture remains even after 60 minutes of continuous establishment of the equilibrium state. If the reagent solution is poured into the side arm of the vessel after this time, there

TABLE 1

Microdetermination of Active Hydrogen by the Manometric Method

Compound	mg Sample	%H Calc.	%H Found	%H Δ
Benzoic acid	1.445	0.825	0.819	-0.006
	1.526		0.821	-0.004
	1.477		0.821	-0.004
	1.487		0.830	+0.005
	1.323		0.829	+0.004
Acetanilide	0.826	0.745	0.753	+0.008
	1.121		0.727	-0.018
	1.782		0.741	-0.004
Pyrogallol	0.587	2.398	2.378	-0.020
	0.549		2.398	0
	0.511		2.386	-0.012
Vanillin	1.962	0.663	0.667	+0.004
Diphenylamine	1.450	0.595	0.605	+0.010
	1.946		0.582	-0.013
p-Toluenesulfonamide	1.032	1.178	1.174	-0.004
	1.026		1.195	+0.017
α-Naphthylamine	0.924	1.408	1.403	-0.005
	0.964		1.422	+0.014
DL-2, 3-Bis(carbo-benzoxyamino) propionamide	0.939	1.085	1.085	0
	1.016		1.073	-0.012
	1.136		1.070	-0.015
Tosyl-L-glutamine	1.197	0.962	0.942	-0.020
	1.012		0.973	+0.011
	1.251		0.956	-0.006

TABLE 1 (CONT'D.)

Compound	mg Sample	%H Calc.	%H Found	Δ
Cyclohexanecar-boxamide	0.932	1.585	1.565	-0.020
α-Terpineol	1.948	0.654	0.662	+0.008
Methycyclohexanol	1.823	0.883	0.903	+0.020
	1.449		0.887	+0.004
Ethanedithiol	0.569	2.140	2.131	-0.009
	0.970		2.130	-0.010
Aniline	0.897	2.165	2.180	+0.015
	0.654		2.162	-0.003
	0.715		2.175	+0.010
1, 3-Dimethyl-4-iso-propyl-3-cyclohexanol	1.920	0.592	0.590	-0.002

is a further pressure increase, which averaged 6 mm of manometer liquid column in the whole series of experiments. The measured pressure increase is therefore corrected by this amount, in the analysis of solid samples only. In the described microdetermination method for active hydrogen, this is the only correction (other than continuous corrections for atmospheric pressure and bath temperature during the experiment) incorporated in the final calculation of the active hydrogen content in the sample. In view of the rapidity of the reaction, the above-mentioned constant pressure increase of 2-3 mm of manometer liquid column in 30 minutes has no practical effect on the accuracy of determination, and can therefore be neglected.

After connection of the vessel with the weighed sample to the manometer and removal to the thermostat bath, a pressure increase takes place. The greatest increase appears within the first minutes, during which time it is caused mainly by evaporation of the tetrahydrofuran in the $LiAlH_4$ solution into the vessel until an equilibrium state is established between the liquid and gas phases, which is a function of the bath temperature, and also by the reaction of trace oxygen and moisture in the vessel with the reagent. Also contributing to the pressure increase is the moisture adsorbed on the walls of the manometer vessel (Fig. 2), whose residue in the side arm makes necessary the only correction in the described determination.

The pressure increase decreases very rapidly with time, and after about one hour of thermostating adjusts to a value of 2-3 mm/30 min, which remains constant for several hours. Since the amount of oxygen and moisture in the closed flask can be regarded as final, and the establishment of equilibrium between tetrahydrofuran and its saturated vapors is also relatively rapid, the observed increase must have another cause. As we shall show below, it is due to spontaneous decomposition of the $LiAlH_4$ and gradual, although very slow, extraction of the lubricant from the ground joint by the tetrahydrofuran vapors and their subsequent reaction with the reagent, with liberation of hydrogen.

The spontaneous decomposition of $LiAlH_4$ (see also [24]) is confirmed by the following lengthy experiment. In order to exclude all the other cited factors causing pressure increase, an all-glass manometer was used, provided with sealed-on flasks instead of stop-cocks and ground stoppers. During cooling, dry nitrogen was passed through the empty manometer side arms from which the liquid had been drawn down into the reservoir, and was let out through the side arm of the vessel. The reagent solution was also pipetted in in this manner, and the side arm was then closed off. The whole system was closed to the atmosphere by forcing the manometer liquid back into the manometer capillaries, and was then placed in the bath. The thermobarometer was prepared in the same manner; this contains only pure tetrahydrofuran instead of the reagent solution in order to eliminate a possible pressure change caused by uptake of tetrahydrofuran vapor by the manometer liquid in the manometer side arm. The pressure increases in the vessel were read regularly every hour only on the second day, so as to ensure not only establishment of equilibrium between tetrahydrofuran and its vapor, but also reaction of total moisture and trace oxygen in the gas phases as well as moisture adsorbed on the walls of the vessel. On the second and third day of this lengthy experiment, the average pressure increase was 0.5 mm of manometer liquid column per hour, this increased to 1 mm/hour on the fourth day and reached a value of 1.5 mm/hour on the fifth. The gradually increasing pressure increase proves the autocatalytic decomposition of $LiAlH_4$.

Another lengthy experiment, using normal Warburg manometers, lubricated with lanolin, was carried out in the same manner. 1 ml of a solution of $LiAlH_4$ in tetrahydrofuran was again pipetted into the manometer vessels, and pure tetrahydrofuran into the thermobarometer vessel. The pressure increase averaged 2 mm/hour on the second day, increased to 3 mm/hour on the third day, and to 4 mm/hour on the fourth. The increased pressure rise in this experiment was caused by extraction of lanolin from the ground joints, which proceeds to such an extent during this time that the water from the bath penetrates into the manometer vessel on the fifth day.

DISCUSSION

The simple, sensitive and accurate Warburg manometer technique, originally developed for biochemical purposes, can be used in a number of cases in which volume changes of very small quantities of the reacting gas must be measured. The advantages of this method–the simplicity of the whole apparatus, the possibility of simultaneous determination of several samples, and the very small consumption of reagents –are also utilized in organic analysis (so far mainly only in the microdetermination of double bonds by catalytic hydrogenation). In our work, the advantages of the Warburg manometric method were also demonstrated for the microdetermination of active hydrogen in organic substances.

The requirement for a successful determination of active hydrogen is a complete dissolving of the analytical sample in the reagent solution. For our 0.5-2-mg samples, the final concentration of the substance being analyzed in the reagent solution was of the order of tenths of one per cent, and one can usually confine oneself to the use of tetrahydrofuran, whose special ability to dissolve the majority of organic compounds is generally known. In exceptional cases, where substances are not soluble in tetrahydrofuran, any other solvent can be used equally well.

In the manometric method, the bath temperature must be kept constant to within $\pm 0.02°$. The fulfillment of this requirement permits use of a solvent with a higher vapor pressure.

In weighing out volatile liquid samples, the capillary must be carefully thermostated before it is sealed off. Any excess or diminished pressure in the capillary can produce a pressure change of several mm after crushing in the manometer vessel. The crushing of the capillary in the manometer vessel proposed by us is in itself very easy and reliable, and does not cause any pressure changes with properly sealed-off capillaries. Substances with a honey-like consistency are weighed into flat weighing glasses or, even better, on to a flat glass fragment. For solid substances, one can equally well use manometer containers whose side arms are provided only with a simple stopper, without the sealed-on rod for crushing capillaries. The weight of the analytical sample is chosen in accordance with the number of active hydrogens and the molecular weight, such that, if possible, the whole manometer scale is used for measuring the pressure increase of the hydrogen liberated.

During the period of 60-90 min, which is indispensably necessary, in the manometric determination of active hydrogen, to establish equilibrium in the whole system [9, 20], the pressure increase drops to 2-3 mm/30 min in the method proposed by us, compared to values three or four times as large in the previously proposed procedures. [20] This very small, constant pressure increase was achieved by us by using a smaller manometer vessel, in which, because of the smaller inner surface of the vessel, less moisture is adsorbed, and the content of moisture and oxygen in the atmosphere of the vessel is thus decreased. Since, in this manner, one also works without preliminary solution of the analytical sample in an inert solvent, from which also traces of moisture diffuse [20], one can rinse almost the whole interior of the vessel with $LiAlH_4$ solution in the analysis of a liquid, and can thus remove the moisture adsorbed on the walls.

The results obtained with a wide variety of substances (see Table 1) prove the high accuracy of the method described, which is often higher than that of the procedures cited in the literature. [18, 25, 26, 27,]

SUMMARY

A simple manometric method for the microdetermination of active hydrogen in solid and liquid organic substances, using a Warburg manometer with a modified vessel, suitable for crushing capillaries, has been developed. The sample weights were between 0.5-2 mg, and the deviations of results from theory did not exceed ± 0.02%.

LITERATURE CITED

[1] T. Zerewitinoff, *Chem. Ber.* **40**, 2023 (1907).

[2] B. Flaschenträger, *Z. Physiol. Chem.* **146**, 219 (1925).

[3] H. Roth, *Mikrochemie* **11**, 141 (1932).

[4] A. Soltys, *ibid.* **20**, 107 (1936).

[5] R. Ciusa, *Gazz. Chim. Ital.* **50**, II, 53 (1920).

[6] H. Gilman and R. E. Fothergill, *J. Am. Chem. Soc.* **49**, 2815 (1927).

[7] A. P. Terent'ev, K. D. Shcherbakova, and N. V. Kremenskaya, *Zhur. Obshch. Khim.* **17**, 100 (1947).

[8] A. P. Terent'ev and N. I. Shor, *ibid.* **17**, 2075 (1947).

[9] J. A. Krynitsky, J. E. Johnson, and H. W. Carhart, *J. Am. Chem. Soc.* **70**, 486 (1948).

[10] F. A. Hochstein, *ibid.* **71**, 305 (1949).

[11] H. Lieb and W. Schöniger, *Mikrochemie ver. Mikrochim. Acta* **35**, 400 (1950).

[12] K. Ziegler and F. Dersch, *Chem. Ber.* **62**, 1833 (1929).

[13] A. Verley and F. Bölsing, *ibid.* **34**, 3354 (1901).

[14] R. J. Williams, *J. Am. Chem. Soc.* **58**, 1819 (1936).

[15] J. F. Eastham and V. F. Raaen, *Anal. Chem.* **31**, 555 (1959).
[16] R. N. Evans, J. E. Davenport, and A. J. Revukas, *Ind. Eng. Chem., Anal. Ed.* **12**, 301 (1940).
[17] W. Schöniger, *Z. Anal. Chem.* **133**, 4 (1951).
[18] M. Souček, *Chem. Listy* **50**, 323 (1956).
[19] M. Jurecek: "Organic Analysis. II," Prague (1957).
[20] W. Lüttgens and E. Negelein, *Biochem. Z.* **269**, 177 (1934).
[21] H. E. Zaugg and B. W. Horrom, *Anal. Chem.* **20**, 1026 (1948).
[22] J. Rudinger and M. Ferles: "Lithium Aluminum Hydride and Related Compounds in Organic Chemistry," Prague (1956).
[23] A. Kleinzeller, J. Málek, and R. Vrba: "Manometric Methods and their Use in Biology and Biochemistry," Prague (1954).
[24] E. Wiberg, R. Bauer, et al., *Z. Naturforsch.* **6b**, 393 (1951).
[25] I. M. McAlpine and P. A. Ongley, *Anal. Chem.* **27**, 55 (1955).
[26] H. C. Brown and O. Hafliger, *ibid.* **26**, 757 (1954).
[27] A. F. Colson, *Analyst* **82**, 358 (1957).

26

*The Use of the Manometric Method for the Microdetermination of Double Bonds in Catalytic Hydrogenation**

J. Horáček and V. Pechanec

The Warburg manometric method has already been used for quantitative microhydrogenation by a number of authors, e. g. [1-3]. The advantage of this method lies in the use of small quantities of the compound being analyzed and in the accuracy of the determination. However, all previous modifications have been limited in most cases to the analysis of solid compounds. The analysis of volatile liquid compounds involves certain difficulties. These compounds must be weighed into glass capillaries, and the capillaries must be crushed in the closed hydrogenation apparatus before the actual determination, which is not always simple. [4, 5]

In the first communication [6] on the determination of active hydrogen by the Warburg manometric procedure, we proposed a simple manometer vessel, which was adapted for the simple crushing of the capillaries containing the weighed sample. We have now used this vessel for the determination of double bonds in organic compounds by quantitative catalytic microhydrogenation.

Collection Czechoslov. Chem. Communs.* **27, 1500-03 (1962)

EXPERIMENTAL PART

Solvents

Analytically pure glacial acetic acid (Lachema) was used without further purification. For compounds insoluble in glacial acetic acid, we have used ethanol or cyclohexane, purified in the usual manner. [7]

Catalyst

Platinum oxide, according to Adams. [8]

Hydrogen

Electrolytic hydrogen was purified by passage through alkaline pyrogallol solution, and dried with anhydrous magnesium perchlorate.

Apparatus

The apparatus used was the same as in the preceding communication. [6] This consists of a Warburg manometer of the usual type with a millimeter scale and a modified manometer vessel. The vessels used have a volume of about 14 ml. Part of the manometer side arm with the vessel is inserted in a water bath, whose temperature is kept constant at $25 \pm 0.02°C$. The contents of the manometer vessels are shaken by reciprocal movement. The manometric liquid is a silicone oil (methylphenylpolysiloxane, $d_4^{20} = 0.973$, viscosity $\eta = 3.637$ centistokes/18.5°), dyed with Sudan yellow.

PROCEDURE

The carefully washed manometer vessel is dried in a drying oven and cooled in a stream of dry nitrogen. Exactly 1 ml of acetic acid or other solvent is pipetted into the bottom of the main part of the vessel, and about 4 mg PtO_2 is added. The weighed sample (0.5-2 mg) is placed in the side arm: solids are introduced in a small glass beaker (height and diameter about 2 mm), compounds with a syrupy consistency on a splinter of flat glass, and volatile liquids in a thin-walled capillary. The vessel is then placed on the lanolin-lubricated ground joint of the manometer, and is secured with spiral springs. A lubricated stopper is inserted lightly into the side arm. The manometric liquid is forced into the upper position, and hydrogen is passed through the manometer via a three-way stopcock at a rate of 10-15 ml/min. The hydrogen leaves the vessel along the stopper that was inserted lightly into the side arm.

After 5 minutes of gas flow, the stopper is pressed into the side arm, then the manometric liquid is drawn off to the lower position, and the whole system is closed off from the atmosphere by turning the three-way stopcock. The hydrogen flow is turned off, and the manometer is placed in the thermostated bath. During the passage of hydrogen through the apparatus, the catalyst is rapidly reduced. However, its complete saturation with hydrogen and the attainment of equilibrium occur only after the manometer has been shaken for 2 hours in the thermostated bath. Then follows the actual determination. The manometric fluid is forced up to the zero mark in the closed arm, and the level in the open arm is marked. The manometer is then removed from the bath, and the crushed capillary is rinsed into the main part of the vessel by repeated tilting. Since a small fraction of the sample solution remains behind in the bulge of the side arm, care must be taken that part of the catalyst gets into the bulge. The manometer is then returned to the bath, and shaking is started. Hydrogenation is very rapid, and is generally complete after 20 to 30 minutes.

The number of double bonds is calculated from the formula

$$\frac{M \cdot h \cdot k}{22{,}412} \text{ (mg sample)}$$

where M is the molecular weight of the sample, h is the pressure change in mm of manometric-liquid column, corrected from thermobarometric data, and k is the flask constant. [9] Some results are shown in Table 1.

DISCUSSION

Hydrogen is best passed through the apparatus for 5 minutes, as indicated in the experimental part. Within this time, all traces of oxygen are removed from the apparatus and the catalyst is reduced. Too long a passage of hydrogen would result in a detectable decrease in the acetic acid or other solvent used, i. e., in a change in the volume of the gaseous and liquid phase in the manometer vessel. This would lead to an uncontrollable change in the flask constant and, therefore, in the analytical results.

The course of the hydrogenation is affected by the quantity of catalyst used. The reduction of a large or small amount of catalyst proceeds relatively rapidly, but the time required for complete saturation of the

TABLE 1

Microdetermination of Double Bonds
by Catalytic Hydrogenation.

Compound (mol. wt.)	mg Sample	No. of double bonds Calc.	Found	Diff.
Benzoic acid (122. 12)	0.592	3	3.01	+0.01
	0.575		3.00	0.00
α-Methylcinnamic acid (162.18)	0.597	4	4.01	+0.01
	0.573		3.99	-0.01
1-Undecene (154.29)	2.376	1	1.01	+0.01
	1.926		0.98	-0.02
3-Ethyl-2-pentene (98.11)	1.602	1	0.98	-0.02
	1.814		1.01	+0.01
Benzene (78. 11)	0.557	3	2.99	-0.01
	0.422		3.00	0.00
2, 6, 11, 15-Tetramethyl-2, 6, 10, 14-hexadecatetraene (274.49)	0.980	4	3.96	-0.04
	1.198		3.97	-0.03

catalyst with hydrogen increases with the amount of catalyst used. We have therefore selected a catalyst quantity that would be completely saturated with hydrogen relatively rapidly and, at the same time, would permit rapid hydrogenation of the analytical sample. In our work we have used 2 mg of sample and 4 mg of catalyst; complete saturation, in agreement with the literature [10], then takes about 90-120 minutes. At the end of this time, one proceeds with the actual analysis, since the pressure drop is only 1 mm of manometric-liquid column in 30 minutes, and therefore cannot influence the result of the analysis. With the arrangement described, the actual hydrogenation of the analytical sample takes a maximum of 30 minutes when acetic acid is used as solvent, or about 40 minutes if the hydrogenation is carried out in ethanol or cyclohexane.

However, small changes in barometric pressure may occur during the actual hydrogenation, which, in view of the high sensitivity of our set-up (manometric liquid with low specific gravity, flask constant about 1.3), will result in an approximately 14-fold pressure change in the apparatus. A thermobarometer is therefore used, which also corrects any possible changes in the bath temperature.

As manometric liquid, silicone oil is used to advantage; it permits us to use the same apparatus both for hydrogenation and for determination of active hydrogen. [6]

The use of a manometer vessel with a device for the simple crushing of capillaries containing weighed liquid samples permits using the advantages of the Warburg manometer method for even the microdetermination of double bonds in volatile liquid compounds. This is demonstrated by the results shown in Table 1. The errors of the determination are within the usual limits of ± 1-2%.

LITERATURE CITED

[1] O. Warburg and W. Christian, *Biochem. Z.* **266**, 387 (1933).
[2] R. Kuhn and E. F. Möller, *Angew. Chem.* **47**, 145 (1934).
[3] H. Willstaedt, *Chem. Ber.* **68**, 333 (1935).
[4] M. Jureček: "Organic Analysis. II", Prague (1957), p. 271.

[5] R. M. Engelbrecht, *Anal. Chem.* **29**, 1556 (1957).
[6] V. Pechanec and J. Horáček, *Coll. Czech. Chem. Commun.* **26**, 1805 (1961).
[7] V. Herout, B. Keil, et al.: "Laboratory Technique in Organic Chemistry", Prague (1954), pp. 640, 651.
[8] R. Adams and R. L. Shriner, *J. Am. Chem. Soc.* **45**, 2171 (1923).
[9] A. Kleinzeller, J. Málek, and R. Vrba: "Manometric Methods and their Use in Biology and Biochemistry", Prague (1954), p. 13.
[10] H. Roth, et al., in: "Methoden der organischen Chemie (Houben-Weyl)", 4th Ed., Stuttgart (1953), v. II, p. 299.

27

*The Microdetermination of Active Hydrogen Using Gas Chromatography**

M. N. Chumachenko
and
L. B. Tverdyukova

A number of methods have been proposed for the determination of active hydrogen in organic compounds, based upon the reaction of organic compounds with Grignard reagents [1-8] or $LiAlH_4$ [9-11], on titration [12], and acylation [13]. In the methods based on reaction with CH_3MgI and $LiAlH_4$, the quantitatively liberated CH_4 or H_2 is determined gasometrically [1-11], manometrically [14, 15], or by isotopic analysis, using $LiAlD_4$ or $LiAlT_4$ [16, 17]. Some authors have proposed combustion of the liberated gases and gravimetric [18] or titrimetric [19] determination of CO_2 and H_2O. The desire to exclude a blank test led to the development of methods in which the compound to be analyzed reacts directly with the reagent solution, without first being dissolved in an inert solvent. [9, 15, 20] The apparatus used in these methods is in most cases a modification of the original, simple Zerewitinoff apparatus, and has the same disadvantages in that the analytical results are affected by fluctuations in temperature and pressure.

We have developed for the first time a micro

Doklady Akad. Nauk SSSR* **142, 612-14 (1962)

method for determining active hydrogen in organic compounds, with a chromatographic determination of the amount of gas liberated in the reaction of the compound with a solution of $LiAlH_4$ in tetrahydrofuran. The choice of a chromatographic method was dictated by the need for separating hydrogen and solvent vapors on the elution diagram and by the characteristics of the reaction. This method makes it possible to obtain accurate results and to decrease the time of one determination to 5-7 minutes. The accuracy of our method is ensured by the sensitivity of the detector, whose action is based on the measurement of the thermal conductivity of gases, thus eliminating the effect of temperature and pressure fluctuations on the analytical results; by the great difference in the thermal conductivity of hydrogen and nitrogen, which acts as the carrier gas; and by the elimination of a blank test (the reaction is carried out directly with the solvent solution, without first dissolving the sample, and the amount of hydrogen is determined graphically from a calibration curve established with a known substance under the conditions of the analysis).

We constructed an apparatus (Fig. 1) consisting of a reaction system and a chromatograph provided with a special device which makes it possible to carry out the reaction directly in the chromatograph system, and to conduct all the gaseous reaction products into the detector. This is accomplished by means of a pneumatic flow switch consisting of four chambers separated by three rubber diaphragms. Control of the flow switch is accomplished through a four-way valve. Depending upon the position of the diaphragm switch, the carrier gas goes directly into the sensing cell of the detector or goes first through the reaction flask. The most suitable of all the reaction systems tested by us was that shown in Fig. 2. The reaction flask, having capillary outlets at the same height, has a small volume, with no dead space, and is flushed rapidly and thoroughly by the stream of carrier gas. A funnel is connected to the flask by means of a closable ground glass joint. The residue of reagent acts as a seal. Dry nitrogen is used as carrier gas, the flow rate (55-60 ml/min) being kept strictly constant by means of a control valve.

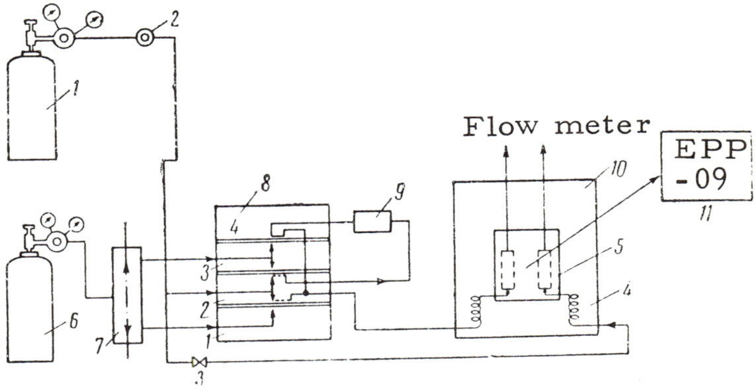

Fig. 1 Apparatus. 1: Carrier-gas cylinder; 2: low-pressure reducing valve; 3: control valve for sensing cell; 4: coil; 5: detector; 6: gas cylinder to actuate diaphragm valve; 7: four-way valve; 8: pneumatic flow switch; 9: reaction system; 10: thermostat; 11: EPP-09 electronic potentiograph.

Investigation showed that the most suitable chromatographic column for our purposes is a coil of metal tubing (length 2 m; i. d. 5 mm) packed with BAU-2 activated charcoal and placed in a thermostat containing distilled water at room temperature; the best solvent is apparently absolute tetrahydrofuran, because of its chemical inertness and high solvent power. However, in each individual case, a better solvent for the analytical sample can be selected from among those that are sufficiently good solvents for $LiAlH_4$. A solution of $LiAlH_4$ in tetrahydrofuran is prepared by shaking 3 g of finely ground $LiAlH_4$ in 100 ml of absolute tetrahydrofuran, followed by decantation. Because of the careful selection of all conditions of the determination, the peaks obtained in the elution

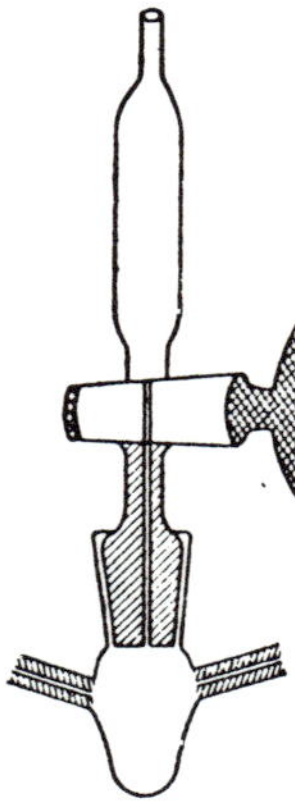

Fig. 2

diagram are symmetrical in form (Fig. 3). This permits calculation of their area by the usual formula for determining the areas of triangles.

8-Hydroxyquinoline, purified by distillation, was used as a standard for establishing the calibration diagram, using samples between 1.187 and 16.001 mg. The calibration curve shown in Fig. 4 was drawn through the 30 points obtained.

The method was used to analyze a number of

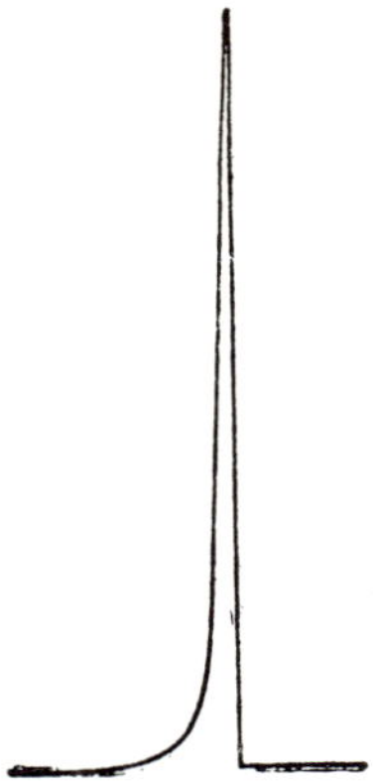

Fig. 3

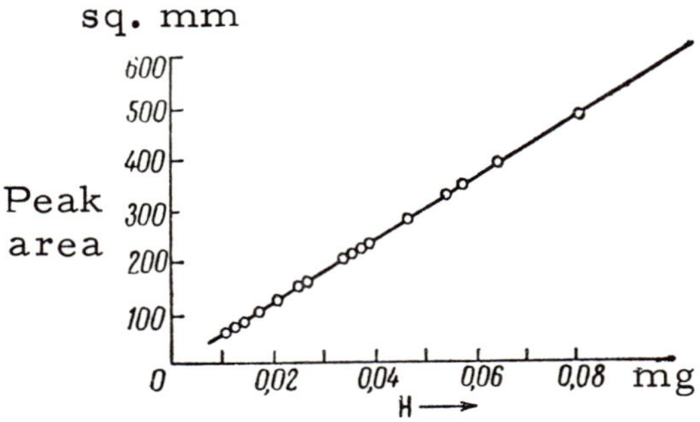

Fig. 4

compounds, for all of which satisfactory results were obtained. Some of these are shown in Table 1. A similar method will be published in *Zhur. Anal. Khim.*

TABLE 1

Sample number	Name and empirical formula of compound	mg Sample	Peak area (sq.mm)	% active H Found	% active H Calc.	Δ
1	3, 15β-Dihydroxy-$\Delta^{1,3,5,(10),8}$-18-nor-D-homoestratetraene-17a-one methyl ether $C_{19}H_{22}O_3$	6.289 4.897	119 101	0.32 0.35	0.34	-0.02 +0.01
2	3, 14ξ-Dihydroxy-$\Delta^{1,3,5,(10),8}$-D-homoestratetraene-17a-one methyl ether $C_{20}H_{24}O_3$	5.760 3.540	111 70	0.32 0.33	0.32	0.00 +0.01
3	17aβ-Ethyl-19-nor-D-homo-testosterone $C_{21}H_{32}O_4$	3.674 3.212	68 64	0.32 0.33	0.32	0.00 +0.01
4	17aβ-Methyl-$\Delta^{1,3,5(10),8}$-D-homoestratetraene-3,17aβ-diol methyl ether $C_{21}H_{28}O_2$	3.153 3.260	63 66	0.33 0.34	0.33	0.00 +0.01
5	$\Delta^{1,3,5,(10)8,14}$-D-Homoestrapentaene-3, 17a β-diol methyl ether $C_{20}H_{23}O_2$	5.325 1.040	116 22	0.36 0.35	0.34	+0.02 +0.01

6	14ξ-Hydroxy-$\Delta^{4,9}$-19-nor-D-homoandrostadiene-3, 17a-dione	4.929	140	0.47	<u>0.44</u>	+0.03
		2.645	73	0.47		+0.03
7	*syn, cis*-1α, 4α-Dihydroxy-$\Delta^{9(14)}$-dodecahydrophenanthren-7-one ethylene ketal $C_{14}H_{20}O_4$	2.752	134	0.81	<u>0.80</u>	+0.01
		2.890	142	0.81		+0.01
8	*threo*-β-(*cis*-2-Hydroxycyclohexyl)-β-dimethylaminoethanol, $C_{10}H_{21}NO_2$	1.801	116	1.05	<u>1.07</u>	-0.02
		1.910	120	1.07		0.00
9	*threo*-β-(*trans*-2-Hydroxycyclohexyl)-β-dimethylaminoethanol $C_{10}H_{21}NO_2$	1.445	89	1.04	<u>1.07</u>	-0.03
		1.753	113	1.07		0.00
10	Benzoic acid $C_7H_6O_2$	1.020	51	0.83	<u>0.83</u>	0.00
		2.238	109	0.81		-0.02
11	N-Trifluoroacetylybenzylamine $C_9H_6F_3NO$	4.090	123	0.50	<u>0.50</u>	0.00
		4.606	146	0.53		+0.03
12	2, 5-Diphenyl-2-pyrazolinone $C_{15}H_{12}N_2O$	4.075	102	0.42	<u>0.43</u>	-0.01
		6.014	155	0.43		0.00

LITERATURE CITED

[1] T. Zerewitinoff, *Chem. Ber.* **40**, 2023 (1907).
[2] B. Flaschenträger, *Z. Physiol. Chem.* **146**, 219 (1925).
[3] H. Roth, *Mikrochemie* **11**, 141 (1932).
[4] A. Soltys, *ibid.* **20**, 107 (1936).
[5] R. Ciusa, *Gazz. Chim. Ital.* **50**, II, 53 (1920).
[6] H. Gilman and R. E. Fothergill, *J. Am. Chem. Soc.* **49**, 2815 (1927).
[7] A. P. Terent'ev, K. D. Shcherbakova, and N. V. Kremenskaya, *Zhur. Obshchei. Khim.* **17**, 100 (1947).
[8] A. P. Terent'ev and N. I. Shor, *ibid.* **17**, 2075 (1947).
[9] J. A. Krynitsky, J. E. Johnson, and H. W. Carhart, *J. Am. Chem. Soc.* **70**, 486 (1948).
[10] F. A. Hochstein, *ibid.* **71**, 305 (1949).
[11] H. Lieb and W. Schöniger, *Mikrochemie ver. Mikrochim. Acta.* **35**, 400 (1950).
[12] K. Ziegler and F. Dersch, *Chem. Ber.* **62**, 1833 (1929).
[13] A. Verley and F. Bölsing, *ibid.* **34**, 3354 (1901).
[14] W. Lüttgens and E. Negelein, *Biochem. Z.* **269**, 177 (1934).
[15] V. Pechanec and J. Horáček, *Collection Czech. Chem. Commun.* **26**, 1805 (1961).
[16] R. J. Williams, *J. Am. Chem. Soc.* **58**, 1819 (1936).
[17] J. F. Eastham and V. F. Raaen, *Anal. Chem.* **31**, 555 (1959).
[18] R. N. Evans, J. E. Davenport, and A. J. Revukas, *Ind. Eng. Chem., Anal. Ed.* **12**, 301 (1940).
[19] W. Schöniger, Z. *Anal. Chem.* **133**, 4 (1951).
[20] M. Souček, *Chem. Listy* **50**, 323 (1956).

28

The Use of Lithium Aluminum Hydride for the Microdetermination of Active Hydrogen*

A.P. Terent'ev, G.G. Larikova
and E.A. Bondarevskaya

Lithium aluminum hydride was discovered relatively recently [1], but has found wide use in synthesis and as an analytical reagent, particularly for determining the content of active hydrogen in organic compounds. [2-12] The determination is carried out in equipment analogous to that used for determinations with CH_3MgI in different variations of the Chugaev-Zerewitinoff method. A weighed sample of the substance is added to a solution of $LiAlH_4$, and the amount of hydrogen liberated is determined from the increase in pressure or volume.

A significant disadvantage of all known methods is the fact that they do not permit the use of highly volatile diethyl ether, but require the use of less volatile solvents (dibutyl, dipropyl, and diisoamyl ethers, etc.), the introduction of a correction for vapor pressure, and the thermostating of the whole apparatus. There has been one attempt to use an ether solution of $LiAlH_4$, but this led to a complication of the method. [2, 7] All of these solvents, as a rule, are poor solvents for di- and polyhydroxy compounds. Such compounds dissolve readily in pyridine, but this is very inconvenient because of its difficult, and often also

---

**Zhur. Anal. Khim.* 18, 514-19 (1963)

dangerous, purification, and because of the need for introducing large corrections. [13, 14] For storage and dispensing, the authors propose sealing the $LiAlH_4$ solution into ampoules, or storing it in special equipment under nitrogen. [2, 3, 4, 7, 9, 11] In all operations, it is necessary to remove moisture carefully from the reaction space and from solvents, and to conduct the reaction in an atmosphere of a specially purified inert gas. The apparatus used is quite complex. The method is used generally in well equipped laboratories of institutes. To make its use more widespread, simpler and more convenient equipment and methods for carrying out the analysis are needed.

In a recently published preliminary communication [15], Chumachenko and Tverdyukova suggested replacing direct measurement of the gas volume in a eudiometer by a considerably more complex gas-chromatographic apparatus. Tetrahydrofuran was proposed as solvent. It is evident that the introduction of other solvents requires special changes. Unfortunately, the authors were apparently not familiar with the micro method for the determination of active hydrogen published in the Soviet literature [16-21], where the analysis is carried out by means of a solution of CH_3MgI in ether and other solvents, and the gas is displaced by vapors of boiling ether. This method is widely known and used in many laboratories. We decided to use this variation for the determination of active hydrogen by means of an ether solution of $LiAlH_4$.

$LiAlH_4$ has a number of advantages: the reaction is carried out at room temperature even for compounds showing steric hindrance, and it is unnecessary to prepare the $LiAlH_4$ specially, since it is manufactured industrially.

In the present paper we describe a simplified variant of the determination for ether-soluble organic compounds, and propose a new method for storing and dispensing an ether solution of $LiAlH_4$; the paper describes the preparation of the reagent from LiH and $AlCl_3$, and also from ready-made $LiAlH_4$.

The essence of the proposed method is as follows: In the reaction of an ether solution of $LiAlH_4$ with an ether solution of a weighed sample of an organic compound, the hydrogen liberated is displaced by vapors of boiling ether into an azotometer filled with a 1:1 water-alcohol mixture, where the ether vapors are absorbed. The hydrogen is transferred from the azotometer to a eudiometer for measurement. There is no need to carry out a control experiment, since the moisture in the solvent is removed from the reaction space before carrying out the reaction.

In carrying out the reaction, it is necessary to make sure that the water-alcohol mixture, which contains ether after the analysis, does not get from the azotometer into the eudiometer; the mixture is dyed with ink, and the capillary passage of the azotometer is washed with distilled water before transferring the gas into the eudiometer.

For prolonged storage of an ether solution of $LiAlH_4$, we propose the use of an iron dispensing tank (Fig. 1), which can be made in any given size. For storing and measuring out a working quantity of reagent, it is convenient to use a 100-ml burette with a $\sqcap$-shaped passage with a plug (Fig. 2), which is connected by rubber tubing to a glass bulb; mercury is used as the sealing fluid. The burette is filled with mercury and connected to the siphon of the dispensing tank; the reagent solution is drawn in by lowering the mercury. In this apparatus, it can be stored for a prolonged time without decomposition. Since 1-2 ml is required for one determination, the available supply is sufficient for 50-70 determinations.

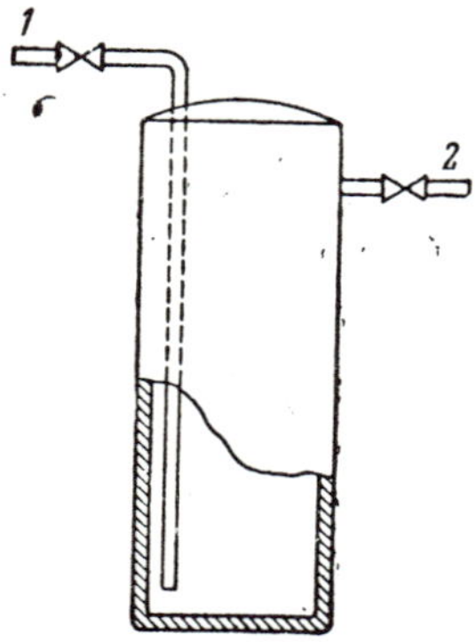

Fig. 1 Iron dispensing tank for storing an ether solution of $LiAlH_4$. 1: Siphon; 2: overflow cut-off valve.

Fig. 3 shows the apparatus for the determination of active hydrogen. The reaction vessel is heated by an electric bulb.

EXPERIMENTAL PART

Reagents

Diethyl ether, dried over calcium chloride and stored over metallic sodium without distillation.

0.25 M solution of $LiAlH_4$ in ether, prepared by a known method. [1] Lithium hydride was ground in a vibratory mill or manually, and was sifted through a 0.5-mm mesh sieve. Manual grinding was carried out in an airtight chamber set up in an exhaust hood. The lithium hydride was analyzed in an apparatus for determining active hydrogen, by hydrolyzing a weighed sample. The activity of the lithium hydride was calculated from the amount of hydrogen liberated. Anhydrous aluminum chloride was used.

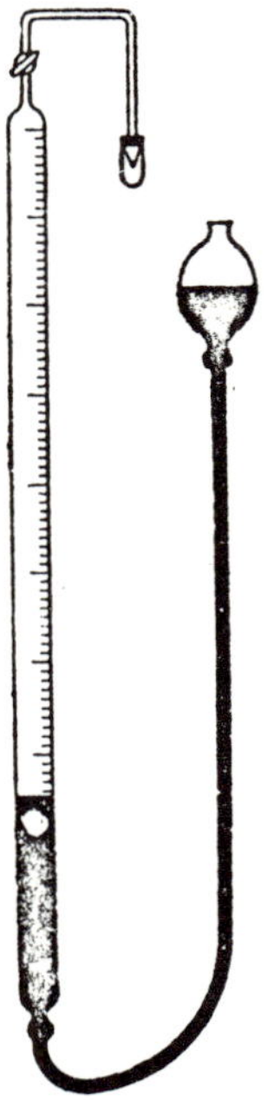

Fig. 2 Apparatus for storing and measuring out a working quantity of reagent.

A 750-ml two-necked flask was mounted in the holder of a magnetic stirrer and provided with a reflux condenser. 6 g (100% activity) of lithium hydride was added to the bottom of the flask through a funnel placed in the second neck. The reaction was seeded as follows: One gram of lithium hydride, 4 ml of ether, and a solution of 1 g of aluminum chloride in 4 ml of ether were placed in a test tube. After visible termination of the reaction, a small portion for seeding was transferred to the flask. Two hundred milliliters of ether was then added, the stirrer was turned

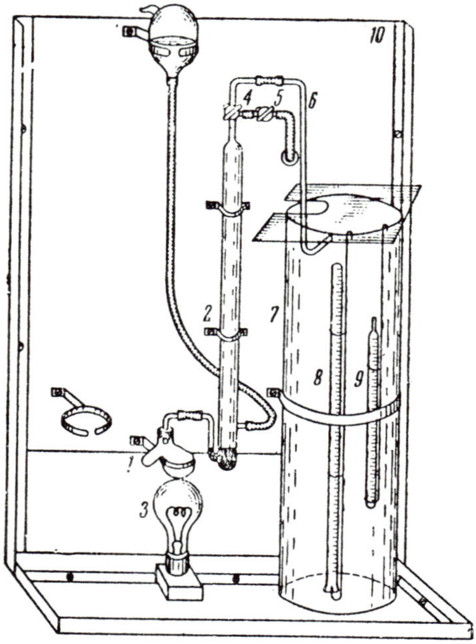

Fig. 3. Apparatus for determining active hydrogen. 1: Glass reactor (capacity about 15 ml), closed with a rubber stopper carrying a tube; 2: azotometer (capacity about 100 ml) with mercury seal and bulb (filled with dyed water-alcohol [1:1] mixture); 3: electric bulb on a support, for heating the reactor; 4: two-way stopcock; 5: simple stopcock; 6: barometer-glass (heavy-wall) tubing; 7: glass cylinder (a large measuring cylinder, capacity 2 l, can be used); 8: eudiometer (capacity 10 ml) with 0.02-ml scale division, attached to an aluminum or copper wire; 9: thermometer with 0.5°graduations; 10: panel for mounting the whole apparatus.

on, and a separatory funnel with a solution of 18 g of aluminum chloride in 300 ml of ether was connected to the flask.

The aluminum chloride was dissolved in the two-necked flask with reflux condenser by stirring. The ether was added dropwise, preventing a strong evolution of heat. The solution obtained, after cooling, was filtered through a glass frit. The aluminum chloride solution from the separatory funnel was added dropwise at a rate such that moderate boiling proceeded in the reaction flask. Stirring was continued throughout the reaction and for 30 minutes after its visible termination. The stirrer was then turned off, and the mixture allowed to settle. The contents of the flask were then poured over into the sampler through a funnel containing a plug of glass wool.

We thus obtained approximately 450 ml of a 0.20-0.25 M solution of $LiAlH_4$; this amount of reagent is sufficient for 200-250 determinations.

To prepare a $LiAlH_4$ solution from the ready-made reagent, a two-necked flask was mounted in the holder of a magnetic stirrer and provided with a reflux condenser. 5 g of $LiAlH_4$ was added to the bottom of the flask through a funnel, and a separatory funnel with 500 ml of ether was attached. Ether was added to the contents of the flask with stirring, the heater of the magnetic stirrer was then turned on, and the mixture was boiled for 1 hour. After cooling and settling, the solution was poured over as described above. An approximately 0.25 M solution was thus obtained.

Determination.

A weighed sample of the compound (5-40 mg, depending on the active-hydrogen content) is taken in a microbeaker and placed in the side arm of the reactor. (The analytical sample must be completely dehydrated, since traces of moisture give a large error.) 1-2 ml of $LiAlH_4$ solution is introduced into the reactor from a burette, and 5 ml of ether, dried over sodium, is added. The reactor is connected to the azotometer, the heater is turned on, and air displaced from the reactor and the connecting tubing into the azotometer by vapors of boiling ether until microbubbles appear. To dissolve the sample, part of the ether vapor is condensed in the side arm, which is cooled with a wad of moist cotton, and heating is stopped.

The air from the azotometer is displaced through stopcocks 4 and 5, stopcock 5 is then closed, and stopcock 4 is turned through 180°. By tilting the reactor, the ether solution of the sample is gradually transferred from the side arm into the reactor. To complete the reaction, careful mixing is carried out by pouring the contents back and forth between the side arm and the reactor. After cessation of the passage of gas bubbles through the mercury seal of the azotometer and a slight movement of the mercury in the connecting tube in the direction of the reactor, the heater is turned on and the hydrogen remaining in the reactor and connecting tube is displaced into the azotometer by ether vapor. After the appearance of microbubbles, the heater is turned off and the reactor disconnected. With stopcock 5 open, the azotometer tube is filled with distilled water drawn in from the cylinder. After 5 minutes, the eudiometer, filled with distilled water, is attached to the end of the exit tube of the azotometer, and the hydrogen is transferred to the eudiometer through stopcock 4. To absorb the alcohol vapors, the eudiometer is shaken by vertical motions and is left at rest for 5 minutes. After this, the gas volume is measured, raising the eudiometer so that the water in it is at the same level as in the cylinder. The temperature and pressure are noted at the same time.

The calculations are carried out by means of the following formulas:

$$\text{Number of H atoms} = \frac{M \cdot V_0}{22{,}380 \cdot a}$$

$$\%H = \frac{1.008 \cdot V_0 \cdot 100}{22{,}380 \cdot a}$$

$$\%OH = \frac{17.008 \cdot V_0 \cdot 100}{22{,}380 \cdot a}$$

where M = molecular weight of the compound; a = sample weight in grams; V_0 = volume of hydrogen liberated, reduced to standard conditions, taking into account the aqueous tension; 1.008 = atomic weight of hydrogen; 17.008 = molecular weight of the hydroxy group.

Table 1 shows the analytical results for ether-soluble compounds, which are in complete agreement with those given in previously published papers. The reactions were carried out at the boiling point of ether in all cases. Reaction with hydroxy compounds and with secondary amines was complete within 1-1.5 minutes. In the analysis of primary amines, the process was complete within 8-10 minutes. A necessary condition for the quantitative course of the reaction for compounds that are poorly soluble in ether, such as butynediol, is the complete solution of the sample in ether, since low results are otherwise obtained. By the example of butyne diacetate it was established that compounds with ester groups react with $LiAlH_4$ without evolution of hydrogen.

It is not necessary to introduce any corrections for a control experiment in this determination, as is also true in the ether variation of the CH_3MgI procedure. [16-23] The analysis, from taking the sample to obtaining the result, requires 20-30 minutes.

Table 1

Compound	mg Sample	Reduced volume of hydrogen (ml)	Number of active hydrogen atoms Found	Calc.
Benzoic acid	15.44	2.82	1.00	1.0
	13.08	2.42	1.01	
	20.00	3.65	1.00	
Salicylic acid	22.97	7.35	1.98	2.0
	22.78	7.36	2.00	
Palmitic acid	31.38	2.70	0.98	1.0
	29.80	2.61	1.01	
	35.53	3.16	1.02	
Stearic acid	39.05	3.10	1.01	1.0
	35.72	2.84	1.01	
1-Naphthol	25.74	4.02	1.01	1.0
	26.55	4.17	1.01	
	22.81	3.59	1.02	
2-Naphthol	22.32	3.47	1.00	1.0
	23.07	3.52	0.99	
	24.14	3.70	0.99	
Phenol	29.42	6.99	1.00	1.0
	29.00	6.95	1.01	
Butynediol	5.31	2.84	2.06	2.0
	5.39	2.86	2.04	
	5.29	2.70	1.96	
Pyro-catechol	30.60	12.26	1.97	2.0
	13.19	4.25	2.00	
8-Hydroxy-quinoline	14.52	2.22	0.99	1.0

Compound	mg Sample	Reduced volume of hydrogen (ml)	Number of active hydrogen atoms Found	Calc.
Aniline	14.56	7.08	2.02	2.0
	14.73	7.38	2.04	
	14.87	7.21	2.01	
1-Naphthylamine	19.53	6.20	2.03	2.0
	20.31	6.29	1.98	
2-Naphthylamine	22.03	6.78	1.97	2.0
	23.27	7.42	2.04	
Diphenylamine	25.25	3.31	0.99	1.0
	28.77	4.08	1.02	
Acetanilide	30.10	4.94	0.99	1.0
	34.20	5.72	1.01	

Preparation of the apparatus for the next experiment takes very little time. The water-alcohol mixture must be heated periodically to remove dissolved ether.

In future communications we shall consider the question of also using other solvents.

SUMMARY

An ether solution of $LiAlH_4$ has been used for determining active hydrogen in organic compounds. A convenient method for storing and dispensing the reagent from a burette over a layer of mercury has been proposed. A simple apparatus for carrying out the analysis, with displacement of hydrogen by ether vapor, has been developed. The method has been used for the analysis of hydroxy and amino compounds.

The determination takes 20-30 minutes. The error is about 3% (relative).

LITERATURE CITED

[1] A. E. Finholt, A. C. Bond, Jr., and H. I. Schlesinger, *J. Am. Chem. Soc.* **69**, 1199 (1947).

[2] J. A. Krynitsky, J. E. Johnson, and H. W. Carhart, *Anal. Chem.* **20**, 311 (1948).

[3] H. E. Zaugg and B. W. Horrom, *Anal. Chem.* **20**, 1026 (1948).

[4] F. A. Hochstein, *J. Am. Chem. Soc.* **71**, 305 (1949).

[5] H. Lieb and W. Schöniger, *Mikrochemie ver. Mikrochim. Acta* **35**, 400 (1950).

[6] W. Schöniger, *Z. Anal. Chem.* **133**, 4 (1951).

[7] J. A. Krynitsky, J. E. Johnson, and H. W. Carhart, *J. Am. Chem. Soc.* **70**, 486 (1948).

[8] G. A. Stenmark and F. T. Weiss, *Anal. Chem.* **28**, 1784 (1956).

[9] A. F Colson, *Analyst* **82**, 358 (1957).

[10] B. Buděšínský, *Collection Czech. Chem. Commun.* **24**, 2948 (1959)

[11] C. R. Dillard, *J. Chem. Educ.* **29**, 129 (1952).

[12] V. Pechanec and J. Horáček, *Collection Czech. Chem. Commun.* **26**, 1805 (1961).

[13] A. Friedrich: "Laboratory Manual of Quantitative Organic Microanalysis" [Russian Ed.], Moscow (1939), p. 213.

[14] "Organic Analysis", ed. J. Mitchell, Jr., et al., New York (1953), vol. I, p. 176.

[15] M. N. Chumachenko and L. B. Tverdyukova, *Dokl. Akad. Nauk SSSR* **142**, 612 (1962).

[16] A. P. Terent'ev, N. I. Avakyan, and

E. A. Terent'eva, *Vestn. Mosk. Univ., Ser. Fiz.-Mat. i Estestven.* Nauk, No. **6**, 99 (1949).

[17] A. I. Kireeva: "Gasometric Microdetermination of the Active Hydrogen Atom by Means of an Ether Solution of Methylmagnesium Iodide (Chugaev Reagent)", Dissertation, Moscow University (1949).

[18] A. P. Terent'ev and A. I. Kireeva, *Izv. Akad. Nauk SSSR, Otd. Khim. Nauk* 1951, 172.

[19] L. Gattermann and H. Wieland: "Laboratory Notes on Organic Chemistry. Supplement to Handbook" [Russian Ed.], Goskhimizdat (1948), p. 96.

[20] S. Veibel: "The Identification of Organic Compounds". Supplement to the Translation, Moscow (1957), p. 317.

[21] N. N. Lezhnev, A. P. Terent'ev, et al., *Kauchuk i Rezina* **20**, No. 11, 21 (1961)

[22] A. P. Terent'ev and N. I. Shor, *Zh. Obsch. Khim.* **17**, 2075 (1947).

[23] N. I. Shor: "The Use of Methylzinc Iodide and Methylmagnesium Iodide for the Determination of the Active Hydrogen Atom," Dissertation, Moscow University (1947).

29

Chromium Sesquioxide as a Combustion Catalyst in the Macrodetermination of Carbon and Hydrogen*

P. N. Fedoseev and M. M. Pavlenko

Many methods for the determination of carbon and hydrogen are described in the literature. The vast majority of these are based on the combustion of organic compounds in a stream of air or oxygen. The combustion products, CO_2 and water, are generally absorbed in special containers. The percentage of carbon and hydrogen in the sample is then calculated from the increase in weight of these containers.

A complete review of the literature on this subject can be found in the detailed monograph by Korshun and Gel'man. [1]

Macro methods for C and H determination have, quite undeservedly, been pushed into the background.

Macro methods for C and H determination are characterized by low cost, availability and simplicity of operation, which have naturally attracted many scientific workers in industrial, educational, and research laboratories, particularly in those cases where a sufficient amount of material is available to the analyst. An excellent explanation of the macro method is given in Men'shutkin's book [2], where the author points out the inadequacy of a number of attempts to

**Zhur. Anal. Khim.* 5, 296-99 (1950)

improve the Liebig method, e.g., replacement of CuO by platinized asbestos and oxygen (Köpfer method), determination of C by a wet method (method of Messinger, Duran, et al), shortened combustion tubes (methods of Köpfer, Dennstedt, et al), etc.

In the present paper, we describe a simple and generally available method for C and H determination in organic compounds, based on the principle of surface combustion. [3]

The principle of the analysis consists in burning the samples in an atmosphere of air or oxygen, in the presence of a surface-combustion catalyst, chromium sesquioxide, at an increased rate of gas flow (100-300 ml/min). Combustion is carried out in hard glass, quartz, or porcelain tubes, of 10-20 mm diameter and length 350-750 mm. The analysis takes 15-20 minutes; 30-40 minutes for highly volatile and sublimable compounds.

The tube is heated by two gas burners, one of which is provided with a horizontal burner pipe, or by electric furnaces. Complete combustion is ensured by supplying a large excess of oxygen to the sample in two channels. (Fig. 1)

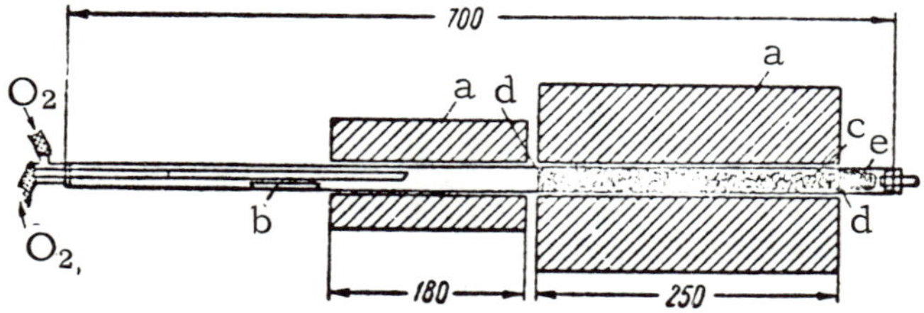

Figure 1

Before starting the experiment, the part of the

tube filled with finely granular Cr_2O_3 (boundary zone) is heated to 600-750°.

After this, the boat with the sample *B* is introduced so that it is 1-2 cm away from the furnace provided for combustion of the sample. The compound is burned by moving the movable electric furnace *A* (or the burner). The sample burns flamelessly on the surface of the catalyst. The combustion process is followed by the movement of the burning layer in the boat. After combustion is complete, the tube is ignited for 5-7 minutes in a stream of oxygen, after which the analysis is considered complete. The absorption train is disconnected and replaced by a $CaCl_2$ tube. The combustion tube can be used for the next analysis immediately after ignition. 5-6 analyses per day can be carried out.

With accurate work, a 100-600 mg sample can be burned without significant decrease in the accuracy of the analysis.

EXPERIMENTAL

Catalyst Preparation

The catalyst required for the tube filling is prepared by heating ammonium bichromate;

$$(NH_4)_2Cr_2O_7 \rightarrow Cr_2O_3 + 4H_2O + N_2$$

The light, flocculent Cr_2O_3 formed, after ignition in air for 1.5-2 hours, is pressed into tablets, which are broken up into pieces 1.5-2 mm in size.

Tube Filling

Hard glass or clear quartz tubes, of 10-20 mm

diameter and 350-750 mm length, are used for the analysis. A small, oxidized copper plug *E*, 2-3 cm in length, is inserted about 4-5 cm from the end of the tube facing the absorption train.

There follows next a small layer of loose, fibrous asbestos *D*, then a 250-300 mm layer of granular Cr_2O_3 *C*, then again a layer of asbestos *D*, which prevents displacement of the Cr_2O_3. In view of the hygroscopicity of Cr_2O_3, the catalyst-filled tube must be ignited in an atmosphere of dry oxygen before starting the analysis.

The weighed sample is placed in the boat as a uniform layer, covering the whole length of the boat. Powdered Cr_2O_3 is carefully poured on top of the sample. If the Cr_2O_3 layer on the sample in the boat is thin, it will ignite with a slight flash or explosion during combustion of the sample, which always leads to a low value for carbon in the sample. Evidently, part of the sample scatters and gets through in an unburned form in such a flash.

For covering the sample, one should have a sufficient amount of dry catalyst in a stoppered test tube. After the sample has been burned and the tube ignited in a stream of oxygen, the Cr_2O_3 from the boat is poured back into the same test tube and is used for subsequent analyses.

Absorption Train

For the absorption train, we have used a Winkler apparatus and U-tubes filled with small pieces of KOH. To retain the fine spray of acid formed when oxygen is passed through rapidly, the large bulb of the coil is half-filled with glass wool. The apparatus must not be filled with acid beyond 3/4 of a turn. Apparatus of too small a diameter should not be used, to avoid encountering resistance to gas flow and decrease in absorptive surface.

Combustion

Before starting combustion, the boundary zone of finely granular catalyst must first be heated to 600-750°. After placing the boat in the tube connected to the absorption train, oxygen is passed through (50-100 ml/min) and the sample is ignited on the side closer to the absorption train. This is done by moving the movable furnace, heated to 600-750°, so that one third to one half of the length of the boat is inside the furnace. As soon as the sample ignites, the stream of oxygen is immediately increased to 250-300 ml/min. The furnace is drawn back, and combustion of the sample in the boat is continued for 2-3 minutes. The combustion is easy to follow, since the catalyst surface is incandescent. After combustion is complete, the stream of oxygen is reduced to its initial velocity and the tube is ignited for 5-7 minutes along its whole length. The method of supplying oxygen to the combustion tube is illustrated in Fig. 2.

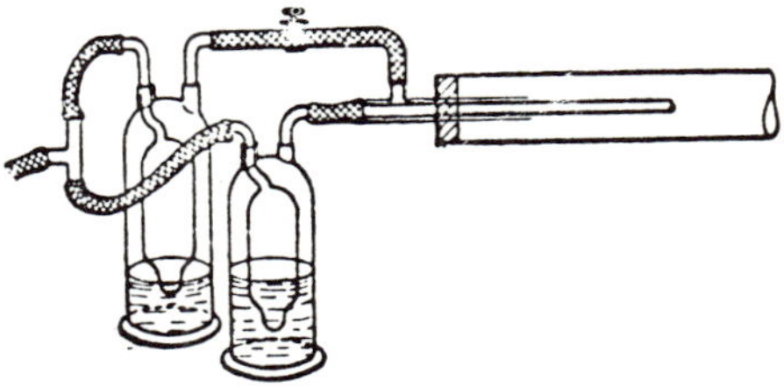

Figure 2

Analysis of highly volatile and sublimable compounds must be carried out with particular care. In

TABLE 1

Sample no.	Compound	Calc. (%)		Found (%)		Δ (%)	
		C	H	C	H	C	H
1	Saccharose	42.08	6.44	42.00	6.51	-0.08	+0.07
				41.93	6.48	-0.15	+0.04
				41.88	6.43	-0.10	-0.01
2	Lactose	42.08	6.44	42.18	6.46	+0.10	+0.02
				42.01	6.49	-0.07	+0.05
				41.98	6.52	-0.10	+0.08
3	Galactose	39.98	6.67	40.00	6.68	+0.02	+0.01
				39.96	6.69	-0.02	+0.02
				39.98	6.72	0.00	+0.05
4	Naphthalene	93.72	6.28	93.55	6.38	-0.17	+0.10
				93.49	6.41	-0.23	+0.13
				93.59	6.39	-0.13	+0.11
5	Benzoic acid	68.86	4.92	68.73	5.11	-0.13	+0.19
				68.62	4.94	-0.24	+0.02
				68.88	5.06	+0.02	+0.14

6	Phthalic anhydride	64.86	2.70	64.92	2.88	+0.06	+0.18
				64.78	2.74	-0.08	+0.04
				64.68	2.75	-0.18	+0.05
7	Succinic acid	40.68	5.08	40.59	5.09	-0.09	+0.01
				40.56	4.98	-0.12	-0.10
				40.52	5.00	-0.16	-0.08
8	Stilbene	93.33	6.67	93.12	6.66	-0.21	-0.01
				93.30	6.81	-0.03	+0.14
9	Retene	92.30	7.70	92.25	7.69	-0.05	-0.01
				92.31	7.84	+0.01	+0.14
				92.27	7.66	-0.03	-0.04
10	Benzene	92.26	7.74	92.13	7.81	-0.13	+0.07
				92.21	7.78	-0.05	+0.04
				92.18	7.75	-0.08	+0.01
11	Toluene	91.24	8.75	91.28	8.56	+0.04	-0.19
				91.11	8.88	-0.13	+0.13
				91.13	8.79	-0.11	+0.04
12	Benzyl alcohol	77.75	7.46	77.60	7.51	-0.15	+0.09
				77.57	7.56	-0.18	+0.10
				77.59	7.42	-0.16	-0.04

TABLE 1 (CONT'D.)

Sample no.	Compound	Calc. (%) C	Calc. (%) H	Found (%) C	Found (%) H	Δ (%) C	Δ (%) H
13	Butanol	64.81	13.59	64.63	13.68	-0.18	+0.09
				64.72	13.63	-0.09	+0.04
				64.79	13.72	-0.02	+0.13
14	Sebacic acid	59.38	8.97	59.41	9.03	+0.03	+0.06
				59.35	9.11	-0.03	+0.14
				59.31	9.06	-0.07	+0.09
15	Phenanthrene-quinone	80.79	3.87	80.76	3.99	-0.03	+0.12
				80.68	4.01	-0.11	+0.14
				80.63	3.92	-0.16	+0.05
16	Fluorenone	86.64	4.47	86.61	4.43	-0.03	-0.04
				86.53	4.56	-0.11	+0.09
				86.47	4.54	-0.17	+0.07

these cases, good results are obtained by dividing the sample uniformly between two boats or putting all in one boat, 10-12 cm in length.

The combustion of liquid samples is carried out in glass ampoules, placed in a boat and covered with catalyst.

The data given in Table 1 indicate the sufficient accuracy of C and H determination in organic compounds (containing C, H, and O) based on the principle of flameless surface combustion.

SUMMARY

1. A method for the rapid determination of carbon and hydrogen in organic compounds is described, based on the principle of flameless surface combustion.

2. Inexpensive and readily available chromium sesquioxide is recommended as catalyst.

3. Increasing the flow rate of oxygen to 350 ml/min does not affect the accuracy of the analysis. At this rate, the reaction products (carbon dioxide and water) are absorbed quantitatively by sulfuric acid and solid potassium hydroxide.

4. The above facts provide a basis for hoping for a very widespread use of this method for determining carbon and hydrogen in various laboratories.

5. Small samples can be analyzed without a significant decrease in the accuracy of the determination.

LITERATURE CITED

[1] M. O. Korshun and N. E. Gel'man: "New Methods of Elemental Microanalysis", Goskhimizdat (1949); M. O. Korshun and V. A. Klimova, *Zhur. Anal. Khim.* 4, 292 (1949).

[2] N. A. Men'shutkin: "Analytical Chemistry", Gosizdat (1928), pp. 392-426; L. Gattermann and H. Wieland: "Laboratory Notes on Organic Chemistry" [Russian Edition], Goskhimizdat (1948), pp. 69-135.

[3] M. B. Ravich: "Surface Combustion", Izdat. Nauk SSSR (1946).

30

Chromium Sesquioxide as a Combustion Catalyst in the Semimicrodetermination of C and H in Compounds with Sulfur and Nitrogen *

P. N. Fedoseev and M. M. Pavlenko

In our earlier communications [1, 2] we have indicated that organic compounds can be analyzed rapidly and accurately, using amorphous chromium sesquioxide, obtained by decomposition of ammonium bichromate, as a catalyst in elemental organic analysis.

It was shown that the analysis can be carried out in an atmosphere of pure oxygen, increasing the flow rate to 350 ml/min, without decreasing the accuracy of the analysis.

We have tried absorbing nitrogen oxides with oxidizing mixtures outside the combustion tube. The apparatus was considerably simplified, and the analysis time was reduced by a factor of 8-10.

In the papers cited, organic compounds containing carbon, hydrogen, oxygen and nitrogen were analyzed.

In the present paper we describe a method of analysis for organic compounds containing sulfur and halogen in addition to carbon, hydrogen, oxygen, and nitrogen.

As a result of the increased oxygen flow rate, passage of halogen and sulfur into the absorption apparatus might have been expected. Our experiments

**Zhur. Anal. Khim*, 8, 158-62 (1953).

have shown that a 10-cm layer of metallic silver (weight approximately 6-7 g) in a 10-12 mm diameter tube completely absorbs sulfur oxides and halogen on heating to 650-750°. The silver may be replaced by lead chromate, but then the length of the layer must be increased to 20 cm.

Some complications are encountered in the analysis of organic compounds containing halogens. In the analysis of these compounds the carbon values, as a rule, are low by 1, 2, or more per cent, depending on the per cent halogen content. The greater the halogen content, the poorer are the analytical results.

Halogens do not affect the accuracy of hydrogen determination. But the following should be noted: If the analysis of halogen-containing organic compounds is carried out with a fresh catalyst, the result obtained will be completely satisfactory. Subsequent analyses give low results for carbon. However, considering that the catalyst is very easy to prepare, inexpensive and available, and that the filling of the tube is not difficult and requires little time, it is possible in individual cases, for low halogen contents, to analyze halogen-containing organic compounds by our method, replacing the catalyst after each determination.

Our method can be used very successfully to analyze coal, petroleum, petroleum residue, oil, and other technical materials.

EXPERIMENTAL

The filling of the tube is carried out as follows. (Fig. 1)

A small asbestos plug is placed 40-50 mm from the end of the tube, then a 100-mm layer of metallic silver and 150 mm of granular chromium sesquioxide, enclosed by asbestos plugs. The silver is used in the form of mesh, twisted into narrow (1-2 mm wide) ribbon, wire, etc.

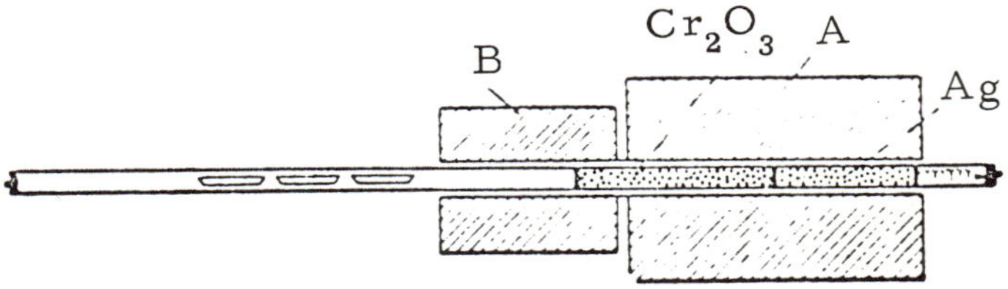

Figure 1.

ANALYTICAL PROCEDURE

The freshly filled combustion tube is ignited for 2-3 hours in a stream of oxygen.

When not operating, the tube is closed with a calcium chloride tube. Before operation is started, a stream of dry oxygen is passed for 10-20 minutes through the tube and the absorption train is connected. The train is then disconnected and transferred to the balance room, where it is kept for 15-20 minutes before weighing.

A 60-300-mg weighed sample is placed in one or several boats, depending on the size of the sample and the nature of the compound. Dry, ignited Cr_2O_3 from a closed test tube is placed on top of the contents of the weighed boats.

Two electric furnaces are used for heating: a large one, *A* (250-300 mm long), and a small one, *B* (150-180 mm long). The temperature of the large furnace is kept constant at 700-750°.

The second, small furnace is used to ignite the

sample and to ignite the empty tube, and must therefore be capable of rapid heating and cooling. This furnace must be movable. The small furnace is turned on after the sample in the boat has been inserted in the tube. After the absorption train has been connected and the sample inserted, the small furnace, heated to 700-800°, is moved. It is moved to a position covering one-third to one-half the length of the boat with the sample, while the oxygen flow rate is increased to 150-200 ml/min. After ignition of the sample, the furnace is moved back and, if necessary (a short, smoky flame), the oxygen flow rate is increased to 250-300 ml/min.

After combustion of the compound in the first boat, the sample in the second boat is ignited in the same way. The combustion time of the sample in the boat is 1-3 minutes, depending on the size of the sample and the nature of the compound. Finally, the tube is ignited for 5-10 minutes in a stream of oxygen, after which the analysis is considered complete. The train is disconnected, weighed, and the carbon and hydrogen percentages are calculated.

Anhydrous $CaCl_2$ is used for absorption of water, and Ascarite or small pieces of KOH for collection of CO_2. For the analysis of compounds containing nitrogen as well as sulfur, one to three coils containing oxidizing mixture and one coil with a saturated solution of a ferrous salt ($FeSO_4$) in 30% H_2SO_4 are placed after the water absorption tube. (Fig. 2)

Concentrated H_2SO_4 must be used to prepare the oxidizing mixture. Concentrated H_2SO_4 is distilled, one third of the distillate is discarded, and the remainder of the distillate is collected and stored in a ground-stoppered flask. After this, a 0.25% solution of $KMnO_4$ in concentrated H_2SO_4 is prepared.

Examples of the analysis of organic compounds,

TABLE I

Sample number	Compound	Calc. (%) C	Calc. (%) H	Found (%) C	Found (%) H	Δ (%) C	Δ (%) H
1	Sulfosalicylic acid	33.09	3.95	33.04	3.99	-0.05	+0.03
				32.97	4.10	-0.12	+0.14
2	*p*-Toluenesulfon-amide	49.11	5.30	49.32	5.31	+0.21	+0.01
				49.27	5.32	+0.16	+0.02
3	Bis(1,1-dimethyl-penten-3-one sulfide)	66.13	8.72	66.26	8.57	+0.13	-0.15
				66.08	8.71	-0.05	-0.01
4	Sulfanilic acid	41.60	4.07	41.53	4.11	-0.07	+0.04
				41.67	4.17	+0.07	+0.10
5	Naphthionic acid	51.71	4.34	51.82	4.39	+0.11	+0.05
				51.79	4.43	+0.08	+0.09
6	Diphenylthiourea	68.39	5.30	68.34	5.26	-0.05	-0.04
				68.38	5.35	-0.01	+0.05
7	Thiosemicarbazide	13.18	5.53	13.36	5.46	+0.18	-0.07
				13.24	5.49	+0.06	-0.04
8	Chlorobenzene*	64.06	4.48	64.03	4.21	-0.03	-0.27
				64.15	4.38	+0.09	-0.10
9	Bromocamphor*	51.96	6.54	51.97	6.32	+0.01	-0.22
				51.71	6.76	-0.25	+0.22
10	1-Chloro-3-nitro-benzene*	45.72	2.54	43.83	2.71	-1.89	+0.17
				42.51	2.50	-3.21	-0.04
11	*p*-Dichlorobenzene*	49.03	2.72	47.12	2.74	-1.91	+0.02
				46.98	2.69	-2.05	-0.04

* Fresh catalyst

TABLE II

Sample number	Fuel series	% Moisture in sample	% Ash in dry fuel	% Total S in dry mass	% Volatile compounds in fuel	% C * VTI	% C ** NKI	% C Δ	% H VTI	% H NKI	% H Δ	No. of analysis check at VTI
1	Chinese coal	0. 59	---	5. 51	14. 71	76. 16	76. 04	-0. 12	3. 56	3. 57	+0. 01	
		0. 59	---	5. 51	14. 71	76. 16	75. 81	-0. 35	3. 56	3. 68	+0. 12	6466
2	Brown coal	6. 56	---	1. 68	39. 59	52. 39	52. 24	-0. 15	3. 27	3. 06	-0. 21	
	dust	6. 56	---	---	39. 59	52. 39	52. 74	+0. 35	3. 27	3. 11	-0. 16	6545
3	Coal	1. 99	---	2. 80	3. 96	74. 91	75. 03	+0. 12	1. 41	1. 33	-0. 08	
	(anthracite)			2. 80	3. 96	74. 91	75. 18	+0. 27	1. 41	1. 38	-0. 03	6491
4	Peat	11. 46	---	0. 38	---	45. 74	46. 08	+0. 34	4. 72	4. 46	-0. 26	
		11. 46	---	0. 38	---	45. 74	45. 85	+0. 11	4. 72	4. 73	+0. 01	5986
5	Intermediate	0. 70	16. 08	---	20. 78	72. 52	72. 50	-0. 02	3. 95	3. 89	-0. 06	5834
	product	0. 70	16. 08	---	20. 78	72. 52	72. 58	+0. 06	3. 95	3. 91	-0. 04	
6	Coal "T"	0. 99	22. 93	2. 93	8. 11	68. 74	68. 57	-0. 17	2. 43	2. 41	-0. 02	
		0. 99	22. 93	2. 93	8. 11	68. 74	68. 36	-0. 38	2. 43	2. 48	+0. 05	5617
7	Coal	0. 59	6. 90	0. 98	14. 61	83. 45	83. 66	+0. 21	4. 07	4. 06	-0. 01	
		0. 59	6. 90	0. 98	14. 61	83. 45	84. 09	+0. 64	4. 07	4. 08	+0. 01	5640
8	Anthracite	2. 83	17. 25	1. 93	3. 88	75. 28	75. 49	+0. 21	1. 32	1. 30	-0. 02	
	"ARSh"	2. 83	17. 25	1. 93	3. 88	75. 28	75. 23	-0. 05	1. 32	1. 34	+0. 02	5644
9	Moscow-	8. 00	34. 04	3. 48	50. 06	40. 99	40. 82	-0. 17	3. 24	3. 16	-0. 08	
	region coal	8. 00	34. 04	3. 48	50. 06	40. 99	40. 79	-0. 20	3. 24	3. 15	-0. 09	5771
10	Silesian	5. 86	18. 82	2. 23	38. 21	59. 00	59. 10	+0. 10	3. 88	3. 70	-0. 18	
	coal	5. 86	18. 82	2. 23	38. 21	59. 00	59. 02	+0. 02	3. 88	3. 79	-0. 09	5753

* F. E. Dzerzhinskiĭ All-Union Scientific Research Institute

** [illegible] Shipbuilding Institute (Nikolaevsk)

coal, and peat, containing sulfur and halogen, are shown in Tables 1 and 2.

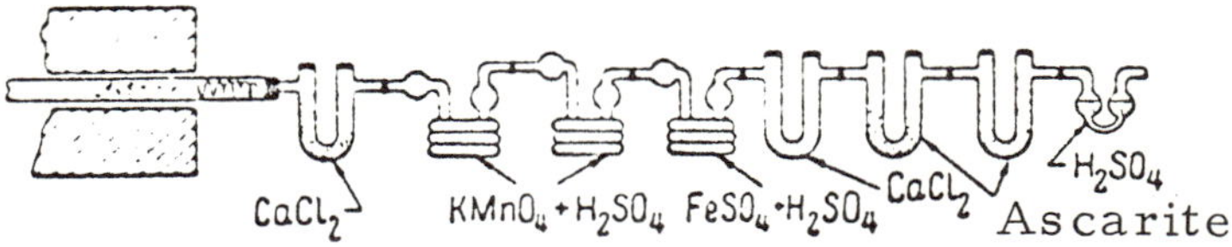

Figure 2.

SUMMARY

1. A method is described for the quantitative determination of carbon and hydrogen in organic compounds containing sulfur and nitrogen.

2. The method was shown to be applicable to the analysis of different forms of organic compounds containing carbon, hydrogen, oxygen, nitrogen, and sulfur.

3. The possibility of using the method for the analysis of various forms of fuel was shown.

4. Using long combustion tubes, large samples can be analyzed by placing them in several boats. This can be very useful in cases where the sample contains very little carbon or hydrogen.

5. The method was considerably simplified and the analysis time was reduced by a factor of 8-10, without decreasing the accuracy of carbon and hydrogen determination.

6. This makes it possible for us to hope for a very very widespread use of the method in research, industrial, and academic laboratories.

LITERATURE CITED

[1] P. N. Fedoseev and M. M. Pavlenko, *Zhur. Anal. Khim.* **5**, 296 (1950).
[2] P. N. Fedoseev and M. M. Pavlenko, *ibid*, **6**, 317 (1951)
[3] M. O. Korshun and N. E. Gel'man: "New Methods of Elemental Microanalysis," Moscow: Goskhimizdat (1949).
[4] M. O. Korshun and V. A. Klimova, *Zhur. Anal. Khim.* **4**, 292 (1949).
[5] M. O. Korshun and V. A. Klimova, *ibid.* **3**, 176 (1948).

31

*Chromium Sesquioxide as a Combustion Catalyst in the Microdetermination of Carbon and Hydrogen in Highly Volatile Liquids**

P. N. Fedoseev and L. S. Ignatenko

We have previously described various procedures and methods used in quantitative organic elemental analysis when working with highly volatile and rapidly decomposing substances [1]. A very successful method in this field is the vacuum combustion of organic substances. However, in order to carry out the analysis in an evacuated tube, the analyst must open the ampoule with the sample immediately after introducing the sample into the combustion tube, and must pump the gas out of it, which requires additional attachments to the apparatus used for the analysis [1, 2].

In the present work, we wanted to find a simpler and, at the same time, more reliable method for the combustion of substances of this kind.

The essence of the method proposed by us consists in combustion without vacuum of a sample of a highly volatile organic liquid in a tube closed on one

* *Izvest. Vysshikh Ucheb. Zavedeniĭ, Khim. i Khim. Tekhnol.* **7**, 797-800 (1964)

side with copper oxide. The sample is contained in an open capillary, which is placed in a quartz capsule with chromium sesquioxide.

ANALYTICAL PROCEDURE

The apparatus is connected in the usual way for the microdetermination of carbon and hydrogen, as described by us previously [3], without including a vacuum pump in the system (see Figure 1).

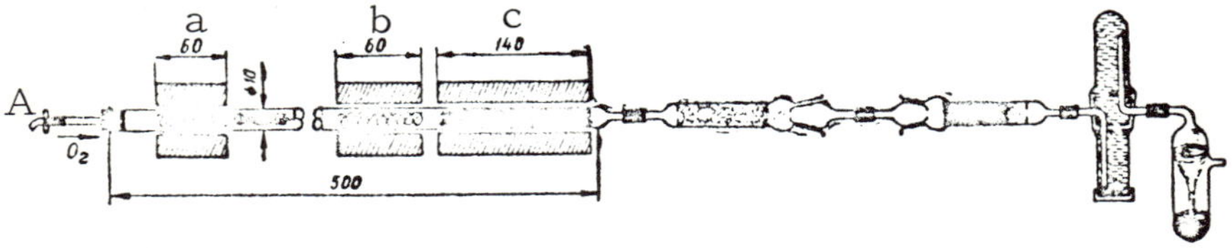

Fig. 1 Apparatus for the carbon-hydrogen determination in highly volatile liquids. *a, b, c*: hinged electric furnaces for microanalysis; A: clamp.

The tube filling remains as before, i.e., a layer of copper oxide and two spirals of oxidized copper. The combustion tube is heated by three hinged electric furnaces. After carrying out blank and check analyses, one proceeds to the combustion of the liquid under investigation.

A 5-8-mg sample is taken in a glass ampoule with a drawn-out capillary (making sure that all of the liquid goes into the body of the ampoule). In sealing off the capillary, its tip is given a hook shape (this form of capillary is convenient for breaking the ampoule). The ampoule is placed in a quartz capsule with compressed chromium sesquioxide. Weighed absorption tubes are attached to the combustion system. Furnaces *a* and *b* have already been brought up to 850-900° and are in readiness behind the combustion tube. Furnace *c* (650-700°) is drawn forward over the copper oxide layer. Clamp *A* is closed, and, before introducing the capsule with the sample into the tube, the hook of the capillary is broken by gently pressing the ampoule against the bottom of the capsule. The capsule (open side toward the absorption tubes) is quickly introduced into the cold part of the tube, to a distance of 100-120 mm from furnace *c*. When burning a low-boiling liquid, it is necessary to cool the part of the combustion tube where the sample is located with ice during the ignition. The small amount of vapor of the liquid which emerges from the capillary is easily adsorbed by the chromium sesquioxide surface, and is held quite firmly by it. Furnace *a* is moved up to the tube (sample in the cold zone!), and, by moving it in the direction of the open end of the tube, the air is displaced (2/3 of its volume leaving through the open end of the tube as a result of the heating). This operation is carried out in 8-12 seconds. The combustion tube is then quickly closed (clamp *A* closed!). Furnace *b* is drawn forward over the tube in position beside furnace *c*. The ice is removed from the part of the tube containing the sample; the capsule is thus between the heated furnaces *a* and *b*. The tube is closed on the oxygen-source side and opened on the absorption-tube side. Furnace *b* (900°) is gradually moved up to the sample (2-3 minutes). The possibility of a flash or explosion is eliminated in

TABLE 1

Analytical Results for Some Samples of Highly Volatile Liquids

mg Sample	H_2O mg	CO_2 mg	%H Calc.	%H Found	%H Δ	%C Calc.	%C Found	%C Δ
Diethyl Ether								
4.720	5.800	11.170	13.59	13.75	+0.16	64.81	64.58	-0.23
7.130	8.801	16.933		13.81	+0.22		64.79	-0.02
Pentane								
4.874	7.307	14.893	16.75	16.77	+0.02	83.24	83.38	+0.14
4.998	7.512	15.275		16.81	+0.06		83.40	+0.16
Acetone								
6.743	6.450	15.391	10.41	10.70	+0.29	62.03	62.28	+0.025
6.587	6.229	15.000		10.58	+0.17		62.14	+0.11

				Hexane				
5.112	7.481	15.677	<u>16.37</u>	16.37	0	<u>83.62</u>	83.69	+0.07
6.017	8.642	18.397		16.07	-0.30		83.43	-0.19
				Methyl Ethyl Ketone				
4.704	4.758	11.490	<u>11.18</u>	11.31	+0.13	<u>66.62</u>	66.65	+0.03
4.624	4.632	11.260		11.20	+0.02		66.45	-0.17
				Cyclohexane				
5.672	7.200	17.833	<u>14.37</u>	14.20	-0.17	<u>85.63</u>	85.80	+0.17
5.941	7.701	18.615		14.50	+0.13		85.50	-0.13
				Hexane				
6.243	8.999	19.139	<u>16.09</u>	16.12	+0.03	<u>83.90</u>	83.66	-0.24
5.977	8.534	18.348		15.97	-0.12		83.77	-0.13
				Isoöctane				
4.934	7.038	15.198	<u>15.88</u>	15.98	+0.08	<u>84.11</u>	84.06	-0.05
4.771	6.723	14.740		15.76	-0.12		84.31	+0.20

view of the small amount of oxygen in the tube, and also by the uniform distribution of the liquid on the amorphous chromium sesquioxide. Combustion of the sample proceeds smoothly, mainly by means of oxygen from the copper oxide present in the combustion tube. The reaction products—water vapor and CO_2—pass into the absorption tubes, as can be observed by the passage of bubbles (of residual sweep gas) through the sulfuric acid bubbler at the exit end of the system. (See Figure 1). After 2-3 minutes, clamp *A* is opened, and a stream of oxygen (40-50 ml-min) is passed through for 4-5 minutes to complete the combustion and to sweep out the reaction products. During this displacement, the combustion tube is ignited by moving the furnace *b*. Sample ignition and sweeping out the reaction products thus takes a total of 6-8 minutes. The absorption tubes are then disconnected and, as usual, are weighed at the determined minute.

We have used the method described for the analysis of a whole series of liquids. Some analytical results are given in Table 1. *

The amorphous chromium sesquioxide for filling the capsule is obtained most simply by decomposition of ammonium bichromate, and is then ignited and stored in a ground-glass-stoppered bottle [4].

Powdered copper oxide was also tested as a capsule filling material but, as shown by experiments, does not retain the vapors of the volatile liquid as well as chromium sesquioxide.

To simplify the drawing of liquid samples with

---

* Compounds of low volatility can be burned without cooling. In these cases, the substance to be analyzed is weighed directly in the quartz capsule, and chromium sesquioxide is sprinkled over the sample.

boiling points below 40-50° (e.g., pentane, diethyl ether, etc.), we suggest a special device: refrigerating forceps; i.e., a hinged, hollow metal chamber filled with pieces of ice (see Figure 2). The same

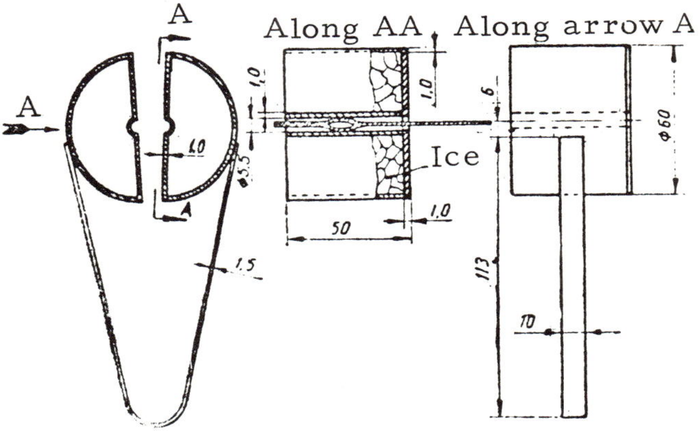

Fig. 2 Device for simplifying the taking of microsamples of highly volatile liquids and cooling the part of the combustion tube in which the sample is located. (The diagram shows the position of the capillary at the moment when the sample of liquid is taken).

device can also be used for cooling the quartz combustion tube with ice while the sample is being introduced.

SUMMARY

A method is proposed for the combustion of highly volatile and rapidly decomposing organic liquids in an open capillary with chromium sesquioxide in a half-closed combustion tube.

The method ensures against flashes and explosions of the sample of highly volatile liquid, and provides sufficient accuracy and reliability of analysis.

LITERATURE CITED

[1] P. N. Fedoseev, L. S. Ignatenko, and T. E. Chernysheva, *Izvest. Vysshikh Ucheb. Zavedenii, Khim. i Khim. Tekhnol.* **1958**, No. 5, 42.

[2] P. N. Fedoseev and L. S. Ignatenko, *Izvest. Akad. Nauk Turkmen. SSR* **1957**, No. 3, 24; **1957**, No. 4, 13.

[3] P. N. Fedoseev and L. S. Ignatenko, *Izvest. Akad. Nauk Turkmen. SSR* **1959**, No. 1, 45.

[4] P. N. Fedoseev and M. M. Pavlenko, *Zhur. Anal. Khim.* **5**, 296 (1950).

32

The Properties of Silver Permanganate and Its Thermal Decomposition Product.

J. Körbl

In experiments on the rapid determination of C and H in organic compounds, our attention was caught by silver permanganate and its thermal decomposition product. Both compounds are strong oxidizing agents. In the case of the decomposition product, we have also found very interesting absorption and catalytic oxidizing properties

Silver permanganate crystallizes in dark crystals with a metallic luster [1], with a lattice that is only slightly deformed in comparision with that of potassium permanganate[2].

On heating, silver permanganate decomposes autocatalytically[3], liberating oxygen. Elovich[4], who studied the decomposition kinetics of solid permanganates, found that the decomposition temperature depends on the cation radius. According to his data, silver permanganate decomposes at 100°.

The decomposition product of silver permanganate is a black, water-insoluble substance. Its composition has been studied by a number of authors. [3, 5-8] According to Grant and Katz[8], the decomposition at 100° in vacuum proceeds in four stages:

* *Collection Czechoslov. Chem. Communs*. 20, 948-51 (1955)

$$AgMnO_4 \rightarrow AgMnO_3 + 1/2\ O_2 \qquad (a)$$

$$AgMnO_3 \rightarrow AgO + MnO_2 \qquad (b)$$

$$2\ AgO \rightarrow Ag_2O + 1/2\ O_2 \qquad (c)$$

$$Ag_2O \rightarrow 2Ag + 1/2\ O_2 \qquad (d)$$

Reactions (a) and (c) go to completion; reactions (b) and (d) are incomplete, so that the decomposition product of silver permanganate consists of x $AgMnO_3$ + y Ag_2O + z Ag. The Debye pattern of the decomposition products shows no interference lines. [6]

A solution of silver permanganate with an excess of silver salts has been used by Hein[9-12] for the determination of gaseous hydrogen. Katz and coworkers[13-17] have used a mixture of silver permanganate and zinc oxide for the determination of small quantities of CO.

This paper describes some preliminary experiments with silver permanganate and its decomposition product. The detailed processing of the results obtained will be the subject of later communications.

EXPERIMENTAL PART AND RESULTS

$AgMnO_4$ is prepared by double decomposition of $KMnO_4$ with $AgNO_3$. [1] The following procedure has proved good: 194 g $KMnO_4$ is dissolved in 4 l boiling water, and the solution is treated with 204 g $AgNO_3$. After dissolving, the solution is allowed to crystallize. The crude product is filtered off with suction on a glass suction filter, washed with 1.5 l water and crystalized from 4 l water. Since concentrated $AgMnO_4$ solution decomposes slightly at 100°, it is advisable to add the crude product to preheated water. After solution, it is filtered rapidly through a glass suction filter, the separated crystals are washed with distilled water and dried at 60-70°. The yield is about 188 g.

Decomposition Product of Silver Permanganate.

About 2 g solid $AgMnO_4$ is heated to decomposition in a test tube. The decomposition product formed has about four times the volume of the original crystals. If larger quantities of $AgMnO_4$ are decomposed, there are some losses due to the scattering of dust. The decomposition product has good permeability for gases, even in high layers, so that it does not have to be granulated for catalytic experiments.

Composition of the Decomposition Product of Silver Permanganate

Manganese is determined as follows in a sample of the decomposition product prepared in different ways:

0. 1 g sample is dissolved in a minimum of conc. HCl, the solution is diluted to 100 ml, and Mn is determined complexometrically after filtering off AgCl.

From the results, the ratio of O to Ag and Mn is calculated. The amounts of Mn and Ag are equivalent.

Preparation of sample	Ratio Ag:Mn:O
Heated in test tube	1:1:2.6
Heated 1 hour at 500°	1:1:2.0
Heated 1/4 hour at 1000°	3:3: 4.0

In the first sample, the lattice is completely destroyed, and the X-ray pattern shows no interference lines. The pattern of the sample heated at 500° shows interference lines after prolonged irradiation [4]; evaluation of the lines requires further experiments. In the third sample, small globules of metallic silver can already be observed, along with a grey-white powdery mass.

Oxidizing Properties of Solid Silver Permanganate

The results of studies on heating compounds with a tenfold excess of $AgMnO_4$ can be summarized as follows:

Organic compounds mixed with $AgMnO_4$ burn from vigorously to explosively. Water and CO_2 are liberated quantitatively from the mixture, while organically bound halogen is retained in the residue in the form of silver halide. Sulfur is converted to Ag_2SO_4 and partly to $MnSO_4$. Pyrites and other sulfides, thiocyanate, and the like react similarly. Sulfates and AgF can be washed out of the remaining residue. Nitrogen-containing organic compounds react to form nitrous gases, part of which is retained as $AgNO_3$ and as traces of $Mn(NO_3)_2$. Ammonium salts, KCN, and other inorganic nitrogen-containing compounds react in the same way. The nitrates formed are decomposed by heating the combustion residue to a higher temperature.

The thermal decomposition product of $AgMnO_4$ reacts in the same way as $AgMnO_4$, but the reaction is less vigorous in this case. When the decomposition product is heated to 500°, its oxidizing power for ammonium salts is lost.

Catalytic Oxidizing Properties of the Decomposition Product of Silver Permanganate.*

The catalytic reactions were carried out in a horizontal glass tube with a 10-mm diameter. One end of the tube is drawn out to a tip, and the tube is charged with a 6-cm layer of the decomposition product.

Catalytic Oxidation of Carbon Monoxide.

A mixture of CO and air (1:20) was passed over the catalyst layer at a rate of 30 ml/min. The reaction was initiated by brief, gentle warming of the initial section of the contact mass with a gas burner. The initial section is maintained at an elevated temperature by the heat liberated during the exothermic reaction. The reaction zone does not advance into the catalyst layer even after several hours of the experiment. The CO_2 formed escapes from the tube.

Catalytic Oxidation of Ammonia.

The gas mixture consisting of oxygen and NH_3 was prepared by passing oxygen through concentrated ammonia solution. After the reaction has been initiated, the surface layer of the contact mass heats up to dull red heat. Nitrogen oxides escape from the tip of the tube, and HNO_3 condenses in the attached receiver. The condensate contained 45% HNO_3.

---

*The catalytic oxidations conducted with the aid of the decomposition product of $AgMnO_4$ are the subject of a patent application. [18]

Absorbing Properties of the Decomposition Product of Silver Permanganate*

It was found that, with the same experimental arrangement as in the preceding experiments, SO_2 is absorbed by the decomposition product at elevated temperatures, forming Ag_2SO_4 and $MnSO_4$. At room temperature, N_2O_4 is preferentially absorbed, in the form of $AgNO_3$, from a gas mixture consisting of CO_2, O_2, and N_2O_4. Preparations moistened with water proved particularly effective.

CONCLUSION

As far as the composition of the decomposition product of silver permanganate is concerned, the oxygen content found by us is in good agreement with that given by Hein[6] and Grant and Katz[8]. The composition of the sample obtained by heating for one hour to 500° could correspond to the reaction sequence proposed by Grant and Katz [8], assuming that reactions (b) and (d) have gone to completion at the elevated temperature, and that the residue consists of silver and MnO_2. However, the experiments of Grant and Katz do not exclude the formation of the compound $2Ag_2O \cdot Mn_2O_3$. The authors arrive at the proposed reaction sequence by a theoretical route. They exclude the possibility of formation of Mn_2O_3 only on the basis of a consideration according to which the formation of Mn_2O_3 by decomposition of silver permanganate would hardly be possible, in view of the lattice of silver permanganate. A decision concerning the

*The catalytic oxidations conducted with the aid of the decomposition product of $AgMnO_4$ are the subject of a patent application. [18]

structure of the "compound" $AgMnO_2$ requires further experiments.

The utilization of the catalytic and absorptive properties of the decomposition product of silver permanganate for the C + H determination will be described in a later communication. The formation of silver halides and Ag_2SO_4 in the oxidation of compounds containing halogen and sulfur could be used analytically for the determination of halogen and sulfur. We are presently working on the determination of sulfur in organic compounds and in pyrites.

SUMMARY

The oxidizing properties of solid silver permanganate and the oxidative-catalytic properties and absorptivity of its thermal decomposition product were studied with a view to their analytical application.

LITERATURE CITED

[1] F. Ullmann: "Enzyklopädie der technischen Chemie", 2nd Ed., Berlin-Wien: Urban & Schwarzenberg (1932), vol. 9.

[2] W. Büssem and K. Herrmann, *Z. Krist.* **74**, 458 (1930).

[3] A. Sieverts and H. Theberath, *Z. Physik. Chem.* **100**, 463 (1922).

[4] S. Yu. Elovich, S. Z. Roginskiĭ, and E. I. Shmuk, *Izv. Akad. Nauk SSSR, Otd. Khim. Nauk* **1950**, 469.

[5] A. Gorgeu, *Bull. Soc. Chim. France* [3] **7**, 261 (1892).

[6] F. Hein, Z. Anorg. Allgem. Chem. **235**, 25 (1937).

[7] E. G. Prout and F. C. Tompkins, *Trans. Faraday Soc.* **42**, 468 (1946).
[8] G. A. Grant and M. Katz, *Can. J. Chem.* **32**, 1068 (1954).
[9] F. Hein and W. Daniel, *Z. Anorg. Allgem. Chem.* **181**, 78 (1929).
[10] F. Hein and W. Daniel, *Chem. Fabrik* **1931**, 381.
[11] F. Hein, W. Daniel, and H. Schwedler, *Z. Anorg. Allgem. Chem.* **233**, 161 (1937).
[12] B. Rassow and L. Wolf, *Chem. Fabrik* **1931**, 407.
[13] M. Katz and S. Halpern, *Ind. Eng. Chem.* **42**, 345 (1950).
[14] M. Katz, G. A. Grant, and R. Riberdy, *Can. J. Research* **28B**, 799 (1950).
[15] G. A. Grant, M. Katz, and R. Riberdy, *Can. J. Technol.* **29**, 511 (1951).
[16] M. Katz, L. G. Wilson, and R. Riberdy, *Can. J. Chem.* **29**, 1059 (1951).
[17] M. Katz, R. Riberdy, and G. A. Grant, *Can. J. Technol.* **30**, 303 (1952).
[18] J. Körbl: "Methods for conducting catalytic oxidations", Czech Patent Application, MZd 89-54.

33

The Thermal Decomposition Product of Silver Permanganate as a Combustion Catalyst for the Carbon-Hydrogen Determination*

J. Körbl

Current combustion methods for C and H determination can be divided into three groups. The first includes methods without a combustion catalyst [1-3]; the second, methods in which a combustion catalyst is used. One of the best known and longest used catalysts is CuO. Ter Meulen [4] used MnO_2 for this purpose, Reihlen [5] worked with Vinosit B, and Ingram [6] carried out the combustion with silver vanadate on pumice or with cerium oxide on Ag_2CrO_4. Fedoseev and Pavlenko [7] recommend Cr_2O_3, prepared by decomposition of $(NH_4)_2Cr_2O_7$. The last group consists of the contact methods [8, 9], based on Dennstedt's principle [10], in which a platinum star heated to a high temperature is used.

The SO_2 contained in the combustion gases and the halogens are usually retained in the combustion tube by silver wool; the nitrogen oxides are fixed by special fillings either in the combustion tube or outside it, between the absorption tubes for H and for C. For this purpose, a whole series of absorption systems has been proposed, e.g., a solution of $K_2Cr_2O_7$ or $KMnO_4$ in H_2SO_4[11], *p*-aminobenzene impregnated with H_3BO_3

**Collection Czechoslov. Chem. Communs.* 20, 953-60 (1955)

and $K_2Cr_2O_7$ solutions [12], silica gel with an adsorbed solution of diphenylamine in H_2SO_4 [13], granulated MnO_2 [14], and, recently, tris(hydroxylamine) phosphate and sulfamic acid [15].

The present paper deals with the use of a contact catalyst obtained by the thermal decomposition of $AgMnO_4$. SO_2 and halogens are fixed directly by the contact mass inside the tube. N_2O_4 is absorbed in an absorption tube filled with the same material, placed between the H and C absorption tubes.

EXPERIMENTAL PART AND RESULTS

REAGENTS

$AgMnO_4$ was prepared in the manner described in the preceding paper [16], and was thermally decomposed in the same way. For absorption of N_2O_4, the decomposition product was moistened with water (5 g is mixed with 1 ml water in a bowl).

APPARATUS

The Pregl apparatus was used [17], after removing the heating mortar required for heating PbO_2. (Fig. 1) The gas burner was adjusted so that the temperature inside the combustion tube was about 450°.

For the absorption of water and CO_2, we have used absorption tubes described in another paper. [18] An absorption tube of the same construction was used to absorb N_2O_4.

Condensation of water in the hydrogen absorption tube is prevented by an electric heater (8 V, 28 ohms), which is energized from the line through a bell transformer. (Fig 2)

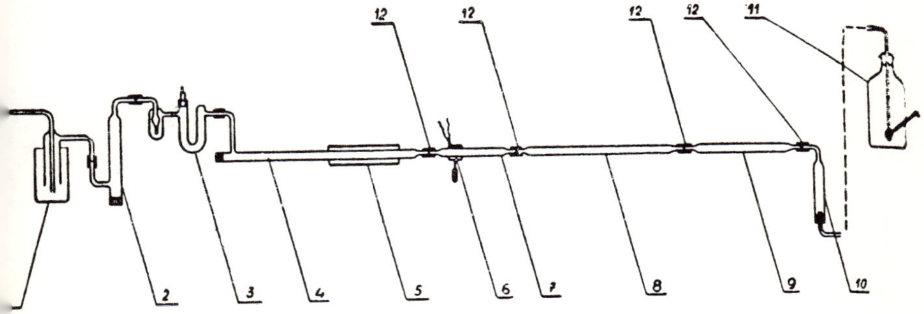

Fig. 1 Apparatus. 1: Pressure regulator; 2: drying tower; 3: bubble counter with U-shaped drying tube; 4: combustion tube; 5: gas furnace; 6: heater (for details, see Fig. 2); 7: H absorption tube; 8: N_2O_4 absorption tube; 9: C absorption tube; 10: $CaCl_2$ tube; 11: Mariotte bottle; 12: rubber connections.

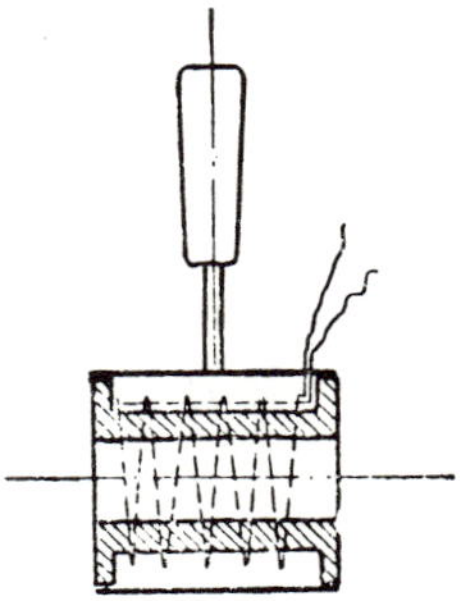

Fig. 2 Heater (part 6 in Fig. 1).

The pressure regulator, drying apparatus, and bubble counter are filled in the usual manner.

Combustion Tube.

A Supremax glass combustion tube (length about 45 cm, o. d. 10 mm) with oxygen inlet side arm is cleaned in the usual manner. Inside the combustion tube at the tip are placed a silver wire (diameter 1 mm) and an approximately 5-cm layer of silver wool, which serves merely to conduct heat*. The silver layer is held in place by means of a small, ignited asbestos plug; however, good permeability of the tube must be ensured. The asbestos layer must not be excessively compressed. About a 15-cm layer of the contact mass obtained by decomposition of $AgMnO_4$ in a test tube is then shaken in. About 4 g $AgMnO_4$ is required to fill the tube. The tube, which is open at both ends, is then placed in the furnace and heated to the operating temperature (450°). The decomposition of $AgMnO_4$, which is accompanied by the evolution of large amounts of oxygen, goes to completion. The decomposition is complete in about 10 minutes. The tube is removed from the furnace, the filling is shaken together well by tapping the sides, and more of the contact mass is added. The catalyst layer is then held in place by a small plug of ignited asbestos, the tube is inserted in the furnace, and the drying apparatus is attached. The permeability of a properly filled

---

*In the analysis of nitrogen-containing compounds, the formation of $AgNO_3$ is observed in cooler parts of the silver zone. This finding is of interest in connection with the question of the decomposition of nitrogen oxides on silver, which was described by Millin [19] and later disputed by Backeberg and Israelstam [20].

tube, at the operating temperature and for an excess oxygen pressure of 5-10 cm water, is about 15-20 ml/min. After rinsing with 250 ml oxygen, the tube is ready for use. As a rule, the first analysis is already correct. After operating for one or two days, the filling is checked and, if necessary, is again shaken together and refilled.

Through the absorption of halogens and sulfur, the structure of the initial layer of catalyst is altered in a sharply defined zone. After about 100 analyses, this zone extends to about 1-2 cm. If the permeability of the tube is reduced thereby, the spent part must be removed and replaced by fresh contact mass. The operation takes only a few minutes to perform.

Absorption Tube for Nitrogen Dioxide.

A tube of the same diameter as the absorption tubes is used. [18] It is filled with Anhydrone (2 cm), then with moist decomposition product (16-20 cm), then again with Anhydrone (5-7 cm). The individual layers are separated by small cotton wads. The absorption tube can be used for 30-40 analyses. Between individual combustions, the tube is kept closed by small pieces of rubber tubing, into which are inserted short pieces of glass rod.

COMBUSTION

About 15 mg of sample is weighed into a platinum boat. The absorption train is attached to the combustion tube by means of rubber tubing. (Fig. 1) First the hydrogen absorption tube with the heater, which is adjusted at the entrance of the absorption tube in such a manner that water cannot condense in the interior, is attached; then the N_2O_4 absorption tube with the shorter Anhydrone layer first; finally, the carbon absorption tube, which is protected by a $CaCl_2$ tube and connected with the Mariotte bottle. The N_2O_4 absorption tube is unnecessary for the analysis of nitrogen-free compounds. If a nitrogen-free compound is burned after one containing nitrogen, use of the absorption tube is recommended, since small amounts of N_2O_4 could possibly remain behind in the apparatus from the preceding analysis, particularly in the hydrogen absorption tube. The boat with sample is placed in the combustion tube, about 3-6 cm in advance of the filling, the oxygen stream in adjusted to 15 ml/min, and the compound is burned in the usual manner to complete the combustion in 10-15 minutes. A decrease in the oxygen stream has no effect on the result of the analysis. Good results were obtained even for compounds that exploded during combustion. After combustion is complete, the apparatus is rinsed with 250 ml oxygen; the oxygen flow rate during rinsing can be increased up to 25 ml/min. The analysis is concluded in the usual manner. A determination takes one hour, including weighing.

DISCUSSION

Despite over a century of development in elemental analysis, only a few oxidation catalysts have been tested in the combustion of organic compounds. CuO, which has been used since Liebig's time, has a large number of disadvantages. The most important is the small active surface of wire-form CuO. In view of the weak catalytic action, one must operate at high temperatures of 750-850° to obtain complete combustion. Other fillings [4-7] have found no application in practice. MnO_2 [4] is, for the most part, irreversibly reduced on contact with vapors of organic compounds, and has only a slight catalytic action. [4, 21] A filling of this type is therefore rapidly exhausted. The modification of Fedoseev and Pavlenko [7] does not give good results in the case of definite compounds; it is recommended by the authors for the analysis of oils, coals, and similar material. The empty-tube method [1-3] requires high temperatures (1000°) and a long combustion space. Korshun [22] carried out oxidation experiments on organic compounds in an empty tube and found that, although the oxidation of hydrogen to water goes practically to completion, part of the carbon is oxidized only to CO. The danger of explosion and the formation of soot are also unpleasant. Methods based on Dennstedt's principle [8, 9] also operate at high temperatures (900-1000°), and, unless care is taken to ensure intimate mixing of the vapors of the compound with excess oxygen at a platinum star, this method has all the disadvantages of the empty-tube combustion method.

The method proposed in this communication has proved itself completely; not only for a series of compounds whose analysis by other methods presents no difficulties, but also for compounds considered hard to burn (steroids), or whose analysis is difficult for other reasons (alkaloids, compounds with high nitrogen content). A series of easily decomposable or explosive substances (e. g., benzoyl peroxide) was also analyzed with equal success. The method described has been used continuously for some time in the analytical department of the Research Institute for Pharmacy and Biochemistry (Prague), and has also been checked in other laboratories. Part of the results obtained are shown in Table 1.

The advantages of the present method can be summarized as follows:

TABLE 1

Compound	Formula	Calc. / Found % C	Calc. / Found % H	Diff. Δ % C	Diff. Δ % H
Biphenyl**	$C_{12}H_{10}$	93.46 93.47	6.54 6.54	+0.01	0.00
Methylvinyl-acetylene*	C_5H_6	90.85 90.61	9.15 9.28	-0.24	+0.13
3-Acetyl-2-undecanone	$C_{13}H_{24}O_2$	73.53 73.51	11.39 11.43	-0.02	+0.04
Diethyl 2, 2-diethyl-4, 5-dioxohexanedioate	$C_{14}H_{22}O_6$	58.72 58.69	7.74 7.90	-0.03	+0.16
Diethyl 3-methyl-4-oxoheptanedioate*	$C_{12}H_{20}O_5$	59.00 59.25	8.25 8.19	+0.25	-0.06
3-Methyl-2-oxo-cyclohexanol	$C_7H_{12}O_2$	65.59 65.66	9.44 9.30	+0.07	-0.14
2-Methyl-2-carbeth-oxy-6-acetoxycyclo-hexanone	$C_{12}H_{18}O_5$	59.49 59.40	7.49 7.60	-0.09	+0.11

1, 1-Bis(*p*-methoxyphenyl)cyclohexane	$C_{20}H_{24}O_2$	81.04 (underlined)	8.16 (underlined)		
		81.25	8.17	+0.21	+0.01
		81.11	8.12	+0.06	-0.04
Diethyl[(2-methoxy-5-bromophenyl)acetyl]-malonic acid	$C_{16}H_{19}BrO_6$	49.62 (underlined)	4.94 (underlined)		
		49.32	5.01	-0.30	+0.07
Ethyl 5-methoxy-1-indanylideneacetate	$C_{14}H_{16}O_3$	72.39 (underlined)	6.94 (underlined)		
		72.35	6.93	-0.04	-0.01
Benzoic acid	$C_7H_6O_2$	68.85 (underlined)	4.95 (underlined)		
		68.80	4.95	-0.05	0.00
		**68.76	4.82	-0.09	-0.03
Benzoyl peroxide	$C_{14}H_{10}O_4$	69.42 (underlined)	4.16 (underlined)		
		69.62	4.40	+0.20	+0.24
Cholesterol	$C_{27}H_{46}O$	83.86 (underlined)	11.99 (underlined)		
		84.11	12.00	+0.25	+0.01
Epidehydroandrosterone	$C_{19}H_{28}O_2$	79.12 (underlined)	9.79 (underlined)		
		79.09	9.76	-0.03	-0.03
3β, 16α-diacetoxy-17α-bromoandrostane**	$C_{23}H_{35}BrO_2$	60.65 (underlined)	7.75 (underlined)		
		60.65	7.85	0.00	+0.10

TABLE 1 (CONT'D.)

Compound	Formula	Calc. / Found % C	Calc. / Found % H	Diff. Δ % C	Diff. Δ % H
Acetamidoacetaldehyde diethyl acetal	$C_8H_{17}NO_3$	54.83	9.78		
		54.92	9.86	+0.09	+0.08
threo-1, 2-Diphenyl-1-(acetylthio)-2-benzamidoethane**	$C_{23}H_{21}NO_2S$	73.57	5.64		
		73.63	5.84	+0.06	+0.20
N-Methyl-N-butyl-*p*-toluenesulfonamide	$C_{12}H_{19}NO_2S$	59.72	7.93		
		59.43	7.99	-0.29	+0.06
2, 2-Dimethyl-7-nonenanilide*	$C_{17}H_{25}NO$	78.70	9.71		
		78.78	9.96	+0.08	+0.25
Diphenylamine	$C_{12}H_{11}N$	85.17	6.55		
		85.19	6.57	+0.02	+0.02
		**85.15	6.61	-0.02	+0.06
2, 4, 6-Trimethylbenzyl 2-diethylaminoethyl ether·HCl	$C_{16}H_{28}ClNO$	67.23	9.87		
		67.29	10.08	+0.06	+0.21

2, 4, 6-Dimethylbenzyl-2-diethylaminoethyl sulfide	$C_{16}H_{27}NS$	72.41 / 72.38	10.26 / 10.29	-0.03	+0.03
(3-Dimethylaminocyclohexyl)diphenylcarbinol·HCl	$C_{21}H_{28}ClNO$	72.92 / 72.79	8.15 / 8.15	-0.13	0.00
3-Acetoxy-17-cyano-pregna-5, 16-diene	$C_{22}H_{29}NO_2$	77.84 / 77.60	8.61 / 8.79	-0.24	+0.18
2-Cyanovaleric acid benzylidenehydrazide	$C_{13}H_{15}N_3O$	68.10 / 68.03	6.59 / 6.63	-0.07	±0.04
1, 3-Dimethyl-2-propylpyrrolidine*	$C_9H_{19}N$	76.53 / 76.50	13.56 / 13.64	-0.03	+0.08
1, 3, 3-Trimethyl-2-pyrrolidinone *	$C_7H_{13}NO$	66.10 / 66.14	10.30 / 10.44	+0.04	+0.14
1-Methyl-2-phenacylpyridinium iodide*	$C_{14}H_{14}INO$	49.57 / 49.82	4.16 / 4.28	+0.25	+0.12
2-Aminopyridine 2-chromonecarboxylate	$C_{15}H_{12}N_2O_4$	63.38 / 63.53	4.26 / 4.42	+0.15	+0.16

TABLE 1 (CONT'D.)

Compound	Formula	Calc. / Found % C	Calc. / Found % H	Diff. Δ % C	Diff. Δ % H
1, 3, 3-Trimethyl-6-(4-pentenyl)-3, 4-dihydro-2(1*H*)-pyridone*	$C_{13}H_{21}NO$	75.30 / 75.59	10.21 / 10.38	+0.29	+0.17
2-Benzoyl-3-ethyl-indole oxime*	$C_{17}H_{16}N_2O$	77.25 / 77.33	6.10 / 6.31	+0.08	+0.21
Ethyl 6-(2-thioxo-4 imidazolidinyl)-hexanoate	$C_{11}H_{20}N_2O_2S$	54.07 / 54.14	8.25 / 8.47	+0.07	+0.22
2-Sulfanilamido-4-methyl-6-propyl-pyrimidine	$C_{14}H_{18}N_4O_2S$	54.88 / 54.81	5.92 / 5.98	-0.07	+0.06
d(+)-2-Butyl-1-hydroxyisoly-sergamide	$C_{20}H_{25}N_3O_2$	70.76 / 70.94	7.43 / 7.72	+0.20	+0.29
Isolysergic acid hydrazide	$C_{16}H_{18}N_4O$	68.04 / 68.10	6.43 / 6.63	+0.06	+0.20

Monoacetylcevine	$C_{29}H_{45}NO_9$	63.14 / 63.24	8.22 / 8.47	+0.10	+0.25
O-Acetyltetramethyl-holarrhimine*	$C_{27}H_{46}N_2O_2$	75.30 / 75.29	10.76 / 10.82	-0.01	+0.06
2-(*p*-Nitrophenyl)-4-methyl-5-cyclohexenyl-2-oxazoline**	$C_{16}H_{20}N_2O_3$	66.64 / 66.58	6.99 / 7.04	-0.06	+0.05
2, 3-Dimethyl-5-nitro-indole	$C_{10}H_{10}N_2O_2$	63.15 / 63.24	5.30 / 5.26	+0.09	-0.04
2-Butanone 2-(*o*-nitrophenyl)-hydrazone	$C_{10}H_{13}N_3O_2$	57.96 / 57.91	6.32 / 6.49	-0.05	+0.17
7-Methyl-1-azabicyclo[1.2.2]heptane picrolonate*	$C_{17}H_{21}N_5O_5$	54.39 / 54.53	5.64 / 5.78	+0.14	+0.14
1-Dimethylamino-3-diphenylmethylenecyclohexane picrate	$C_{27}H_{28}N_4O_7$	62.30 / 62.25	5.42 / 5.58	-0.05	+0.16
2-(*p*-Methoxybenzylthio)-5-aminopyridine picrate	$C_{19}H_{17}N_5O_8S$	47.99 / 48.09	3.61 / 3.70	+0.10	+0.09

TABLE 1 (CONT'D.)

Compound	Formula	Calc./Found % C	Calc./Found % H	Diff. Δ % C	Diff. Δ % H
1, 1, 4, 7, 7-Pentamethyldiethylenetriamine picrate	$C_{27}H_{32}N_{12}O_{21}$	37.68 37.91	3.74 3.88	+0.23	+0.14

The majority of the compounds listed in the table were prepared in the Organic Synthesis Department of the Research Institute for Pharmacy and Biochemistry. The analyses marked with a single asterisk were performed in the Laboratory for Heterocyclic Compounds (Czechoslovak Academy of Sciences); those marked with a double asterisk in the Institute for Organic Chemistry (Czechoslovak Academy of Sciences). Flaschenträger absorption tubes [21] were used in both places.

1. The preparation of $AgMnO_4$ of constant purity and preparation of the decomposition product are very simple. The filling of the combustion tube is rapid and requires only a fraction of the time required for preparing Pregl's universal filling. The filling can be used immediately after insertion, thus eliminating the time-consuming ignition and carrying out of the initial tests. The filling has a long life. One filling generally is good for over 100 analyses, without special operations, and the replacement of the spent part requires only a short time. This method can also be used to advantage when carrying out occasional analyses, without devoting daily attention to the apparatus.

2. The combustion temperature of the new method is significantly lower than for any procedure used in practice. A temperature of 450° is sufficient, and this need not be strictly maintained. Fluctuations of ± 50° are of no significance for the accuracy of the analysis. The low combustion temperature permits the use of an ordinary tube of high-melting glass, with the tube having a sufficiently long life.

3. The proposed mass is also a complete absorbent for SO_2 and halogens, so that the analysis of nitrogen-free materials can be carried out without further adaptations.

4. The decomposition product also absorbs N_2O_4 at normal temperature, so that it is equivalent to the best known absorbents in this respect.

In conclusion, we deem it proper to mention the possible sources of error. Several of these are also encountered in other modifications of the carbon-hydrogen determination. If the catalyst layer has not been shaken together sufficiently by tapping the sides, low carbon values can be caused by inadequate contact of the vapors of the analytical sample with the catalyst. Errors of this type are extremely rare. Correct results were obtained even in tubes in which the catalyst had deposited to the extent of one third of the tube diameter along almost the whole length of the tube. The lower results are mostly caused by excessively large dimensions of the N_2O_4 absorption tube. In this respect, a larger diameter is more harmful than a greater length. Another source of low values is the inadequate function of Ascarite. When functioning properly, the Ascarite is used up vertically, in a sharply defined zone.* Superficial etching indicates its inadequate function. For excessively rapid combustion, a temporary backward surge of the vapors toward the purifying train may occur, so that, at worst, part of the CO_2 and water is absorbed in the purifier. High carbon values, with proper functioning of the purifying train, are caused by exhaustion of the N_2O_4 absorption tube.

Low hydrogen values can be caused by condensation of the water in the protruding section of the combustion tube. The tube must be attached in such a position that condensation is avoided; the tip of the tube may be wrapped around with copper wire, whose end is inserted into the combustion furnace. However, the protruding section of the combustion tube must not be too short, so that the connection between the tube and the hydrogen tube is not scorched. High hydrogen values, with proper functioning of the purifying train, are caused by absorption of N_2O_4 in the hydrogen absorption tube. Errors of this type have been discussed in detail by Belcher and Ingram. [14] Our observations are in good agreement with their statements. N_2O_4 is absorbed in moist Anhydrone. This difficulty is avoided by using the heater for the hydrogen absorption tube.

---

*Ed. Note: Vertical instead of horizontal absorption tubes are referred to here.

SUMMARY

The thermal decomposition product of silver permanganate is used as a combustion catalyst in the determination of carbon and hydrogen in organic compounds. The filling also fixes halogens and sulfur dioxide. The catalytic combustion, which proceeds at relatively low temperature, is rapid and complete, so that the time required for a semimicroanalysis can be reduced to one hour.

The silver permanganate decomposition product was also used for the absorption of nitrogen dioxide between the hydrogen and carbon absorption tubes.

LITERATURE CITED

[1] I. Marek, *Bull. Soc. Chim. France* [4] **43**, 910 (1928).

[2] A. F. Colson, *Analyst* **73**, 541 (1948).

[3] M. O. Korshun and V. A. Klimova, *Zhur. Anal. Khim.* **2**, 274 (1947).

[4] H. ter Meulen and J. Heslinga: "Neue Methoden der organisch-chemischen Analyse", Leipzig (1927).

[5] H. Reihlen, *Mikrochemie* **23**, 285 (1938).

[6] G. Ingram, *J. Soc. Chem. Ind.* **62**, 175 (1943).

[7] P. N. Fedoseev and M. M. Pavlenko, *Zhur. Anal. Khim.* **8**, 158 (1953).

[8] A. Friedrich, *Mikrochemie* **10**, 329 (1931).

[9] K. Bürger, *Chem. Fabrik* **1940**, 325.

[10] M. Dennstedt and F. Hassler, *Z. Anal. Chem.* **42**, 417 (1903).

[11] P. J. Elving and W. R. McElroy, *Ind. Eng. Chem., Anal. Ed.* **13**, 660 (1941).

[12] A. Dombrowski, *Mikrochemie* **28**, 136 (1940).
[13] I. Irimescu and B. Popescu, *Z. Anal. Chem.* **128**, 185 (1948).
[14] R. Belcher and G. Ingram, *Anal. Chim. Acta* **4**, 401 (1950).
[15] C. K. Cross and G. F. Wright, *Anal. Chem.* **26**, 886 (1954).
[16] J. Körbl, *Chem. Listy* **49**, 858 (1955); *Collection Czech. Chem Commun.* **20**, 948 (1955).
[17] F. Pregl: "Quantitative Organic Microanalysis", 4th English Ed., London (1945).
[18] J. Körbl, *Chem. Listy* **49**, 929 (1955); *Collection Czech. Chem. Commun.* **20**, 993 (1955).
[19] D. Millin, *J. Soc. Chem. Ind.* **58**, 215 (1939).
[20] O. G. Backeberg and S. S. Israelstam, *Anal. Chem.* **24**, 1209 (1952).
[21] M. Jureček: "Organic Analysis", Prague (1950).
[22] M. O. Korshun, *Zhur. Anal. Khim.* 7, 96 (1952).

34

A Short Length of the Decomposition Product of Silver Permanganate as Catalyst for the Semimicro Carbon-Hydrogen Determination*

J. Körbl

Various directions can be observed in the development of C and H determination. First of all, there is the endeavor to speed up the analysis. The transition from semimicro- to microdetermination is also usually motivated more by the desire to speed up the analysis than by the need for managing with minute samples. Smaller samples can be burned much more rapidly, and the differences in the behavior of individual compounds during combustion are no longer as critical, which permits a further speeding up of the process. Another direction of development in elemental analysis leads to automation of the combustion process. By an adjustable speed of displacement of the movable furnace, and possibly by other devices, individual authors [1-7] have attempted to take into account the individual differences in behavior of different compounds during combustion, and thus to reduce the demands on the time and attention of the analyst. The third possible direction of development, i. e., the introduction of more effective catalysts, which would also make possible a more rapid combustion of larger samples, and for which the effects of

*Collection Czechoslov. Chem. Communs. **20**, 1026-31 (1955)

the differences in behavior of different compounds could be neglected, has remained generally neglected up to now.

In the second paper of this series [8], we proposed a new, highly effective catalyst as a combustion tube filling for the semimicrodetermination of carbon and hydrogen, which we prepared by thermal decomposition of $AgMnO_4$. [9] We started with the Pregl apparatus [10], whose dimensions and essential component parts were retained. When using the new contact material, we succeeded in burning all the compounds investigated within 10-15 minutes; the results were within the usual limits of error. The whole determination was thus reduced to one hour.

In the course of further experiments it was found that the catalytic capacity of the decomposition product of $AgMnO_4$ is so great that it is possible to shorten the combustion-tube filling significantly, and thus to eliminate the sources of error associated with large dimensions of the apparatus. The experimental part of this communication deals with the combustion of compounds using a short filling of the decomposition product of $AgMnO_4$, with several simplifications in the apparatus, and with the determination of the capacity of the short filling for the absorption of halogens and SO_2. In one of the later papers it will be shown that the proposed method can also be used for microdeterminations. In view of the fact that the combustion time has already been reduced to a minimum in the semimicrodetermination, no significant saving of time can be expected from the micro method.

EXPERIMENTAL PART AND RESULTS

APPARATUS

Regulating and Purifying Apparatus.

Oxygen was obtained from a steel cylinder through a reducing valve. The pressure regulator, $CaCl_2$ drying tower, and U-tube with bubble counter, filled with Ascarite, Anhydrone, and chromsulfuric acid* were taken over from the Pregl apparatus.

Absorption Train.

As in the preceding paper [8], but without the Mariotte bottle.

Combustion Tube (Fig. 1).

A 25-cm Supremax glass tube (10 mm o.d.) with a 4-cm capillary tip and an oxygen inlet tube sealed on at 5 cm from the end of the tube is used. Into the tip is placed a 1-mm diameter silver wire, which is bent at the inner end so that it cannot fall out. This is followed by about a 2-mm plug of ignited asbestos and about a 3.5-cm layer of the decomposition product of $AgMnO_4$ [8, 9], which is again secured with asbestos. The asbestos plug is compressed so that the transmission rate of the tube is 15 ml/min at the operating temperature of 450-500° and for an excess oxygen pressure of 5-10 cm of water.

---

*Chromsulfuric acid was used instead of the generally used concentrated NaOH. [11]

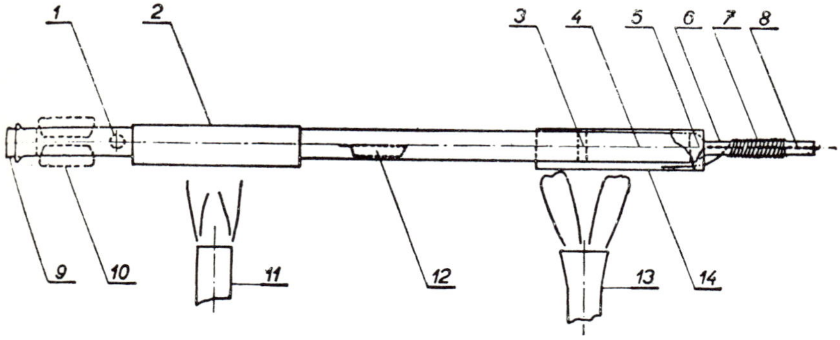

Fig. 1 Combustion apparatus. 1: Oxygen inlet sidearm; 2: brass tube; 3: asbestos layer; 4: layer of $AgMnO_4$ decomposition product; 5: asbestos; 6: tip of tube; 7 and 8: silver wire; 9: rubber stopper; 10: clamp of stand; 11: burner; 12: platinum boat; 13: microburner (details in Fig. 2); 14: covering sheath of the contact filling.

Combustion Arrangement.

The combustion tube is held in a laboratory stand. For indirect heating of the sample during combustion by means of a gas burner, a brass tube (diameter 14 mm, wall thickness 0.33 mm, length 6.5 cm) is used, which is placed over the combustion tube. A second brass tube with the same dimensions serves as a sheath for the contact filling. It is placed over the catalyst layer so that its beginning projects about 1.5 cm beyond the catalyst layer. To prevent condensation of water in the tip of the combustion tube during analysis, about 2 cm of the central part of the tip is wound with 1 mm-diameter copper (or, better, silver) wire, and one end of the wire is inserted under the brass protective sheath. The catalytic filling is heated with a two-flame microburner, which is prepared, e.g., from a Teclu burner by reshaping the originally circular orifice into a two-slitted one by compression. (Fig. 2) The burner is placed so that one of its flames

Fig. 2 Orifice of the microburner.

heats the front end, and the other the central part, of the brass tube. In this way, a suitable temperature gradient is obtained, and the separation of decomposition products of the burning samples before entering the contact layer is prevented. The flame height is so adjusted that the interior temperature of the combustion tube is about 450-500°C (the temperature need be measured only once).

ANALYTICAL PROCEDURE

After assembling the apparatus, the catalytic filling is heated to the operating temperature with the microburner, and the combustion tube is briefly ignited. The oxygen flow is adjusted to 15 ml per minute by lowering or raising the pressure regulator. (The flow rate is controlled by means of the bubble counter, which has been previously calibrated with the aid of the Mariotte bottle or some other suitable device).

After attaching the absorption train, the transmission rate should drop by no more than one third. If the decrease in transmission rate is greater, the oxygen stream can be pulled through by moderate suction with a water-jet vacuum pump. The suction is controlled by means of a screw clamp.

About 15 mg of sample is weighed out, and burned within 5-10 minutes; the combustion tube is then rapidly ignited. After about another 10 minutes of operation, the apparatus is sufficiently rinsed. The absorption tubes are detached and weighed after 15 minutes.

RESULTS

COMPOUNDS CONTAINING C, H, N (O)

For compounds containing C, H (O), the absorption train consists of the C and H absorption tubes [12] and a $CaCl_2$ tube. If water condenses in the capillary connection of the H-tube, the connection is warmed by means of a suitably bent, thick copper wire, one end of which is inserted under the protective sheath of the catalytic filling. The results are shown in Table 1.

In the analysis of compounds containing nitrogen, a N_2O_4 absorption tube [8] was inserted between the H and C absorption tubes. The H tube was heated with an electric heater.[8] Table 2 contains some consecutive analyses of standards.

TABLE 1

Compound	Formula	Calc. Found %C	%H	Δ %C	%H
Benzoic acid	$C_7H_6O_2$	68.85	4.95		
		68.93	4.94	+0.08	-0.01
		68.85	5.04	0.00	+0.09
Benzil	$C_{14}H_{10}O_2$	79.98	4.79		
		80.01	5.03	+0.03	+0.24
Cholesterol	$C_{27}H_{46}O$	83.86	11.99		
		84.02	11.85	+0.16	-0.14
5-Methoxy-1-indanyl-acetic acid	$C_{12}H_{14}O_3$	69.88	6.84		
		69.96	6.91	+0.08	+0.07

TABLE 2

Compound	Formula	Calc. Found %C	%H	Δ %C	%H
Diphenylamine	$C_{12}H_{11}N$	85.17	6.55		
		85.11	6.44	-0.06	-0.11
Urea	CH_4N_2O	20.00	6.71		
		20.23	6.74	+0.23	+0.03
Acetanilide	C_8H_9NO	71.09	6.71		
		71.09	6.57	0.00	-0.14
Benzanilide	$C_{13}H_{11}NO$	79.16	5.62		
		79.31	5.73	0.15	+0.14
Azobenzene	$C_{12}H_{10}N_2$	79.03	5.53		
		79.23	5.56	+0.14	+0.03

COMPOUNDS CONTAINING SULFUR AND HALOGEN

The analyses were carried out in the same apparatus as the analyses of compounds containing C and H (O). The results are shown in Table 3.

TABLE 3

Compound	Formula	Calc. / Found %C	Calc. / Found %H	Δ %C	Δ %H
(Ethylsulfonyl) diisopropylidenefructose	$C_{14}H_{24}O_8S$	47.71 / 47.48	6.86 / 6.83	-0.23	-0.03
6-Bromo-5-hydroxytrimellitic acid	$C_9H_5BrO_7$	35.43 / 35.26	1.65 / 1.73	-0.17	+0.08
{2-[(Phenylcyclohexyloxy) acetoxy]ethyl}-dimethylsulfonium iodide*	$C_{18}H_{27}IO_3S$	48.00 / 48.29	6.04 / 6.11	+0.29	+0.07

*With the long filling, [8] we found C 48.42%, H 6.14%.

DETERMINATION OF THE CAPACITY OF THE FILLING FOR SULFUR AND HALOGEN.

In order to obtain an idea of the absorptive capacity of the catalytic filling, a series of capacity tests was conducted.

Sulfur.

Several consecutive samples of elemental sulfur were burned in a combustion tube with a fresh filling (0.8 g). The increase in weight of the H and C absorption tubes was determined (Table 4).

TABLE 4

	Δ mg	
mg S	H tube	C tube
51.48	+0.18	+0.13
14.14	-0.13	-0.01
15.10	+0.15	+0.70
15.84	+0.04	+27.23

From these results, the absorptive capacity of the filling was estimated at 80 mg of sulfur. Because of reaction with SO_2, the filling cakes and shrinks strongly. The initial black color changes to grey. After combustion of the first sulfur sample, the shrinkage has progressed halfway along the filling.

Chlorine.

Tests were carried out with concentrated HCl. The tip of the freshly filled combustion tube was connected to a bend of glass tubing, one end of which dipped to the bottom of a test tube filled with 5 ml water and a few drops of Tashiro indicator. Tenth-milliliter samples of HCl solution were burned, corresponding to 34 mg Cl. During the first two combustions, there was no change in the indicator color, but in the final phase of the combustion of the third sample, the indicator was bleached and the odor of chlorine could be detected. About 0.08 mg Cl (as HCl) was determined titrimetrically in the absorption receiver. During combustion of the fourth sample, the chlorine odor was already very intense, and upon completion of combustion the receiver contained 11.6 mg Cl (as HCl). The capacity of the filling for Cl is thus about 100 mg. No striking changes are observed in the spent catalyst: there is no shrinkage, and the black color remains intact.

Iodine.

Elemental iodine was "burned" using the same apparatus as in the preceding case. The absorption receiver was filled with starch solution containing a small amount of KI. In the case of the first combustion (108 mg iodine), the absorption tube remained unchanged. At the end of the second combustion (108 mg iodine again), it suddenly turned blue. After combustion was complete, 5 mg iodine was determined titrimetrically in the tube. The filling retained about 210 mg of iodine. As in the previous case, the catalyst shows no macroscopic changes.

Bromine.

The capacity for Br was not determined, since it can be expected that Br will behave like Cl and I.

Fluorine.

For the test experiments we used 4, 4'-difluorobiphenyl, which had previously been analyzed with a long filling [8] (Analysis 1, Table 5). In two out of a total of three analyses, usable C values were obtained with the short filling, while the H values were high by more than 1% (see Table 5). Etching was noticeable in the H tube.

TABLE 5

Calc. for $C_{12}H_8F_2$ (190.2): C 75.78%, H 4.24%

Analysis	Found %C	Found %H	Diff. Δ%C	Diff. Δ%H
1*	75.68	4.44	-0.10	+0.20
2	76.10	5.94	+0.32	+1.70
3	76.00	5.88	+0.22	+1.64
4	75.98	5.61	+0.20	+1.37

*The analysis was carried out by E. Knižáková, of the Laboratory for Heterocyclic Compounds, Czechoslovak Academy of Sciences.

CONCLUSIONS

It has been demonstrated that a short catalytic layer of the thermal decomposition product of $AgMnO_4$ is sufficient for the semimicro combustion of compounds at relatively low temperature (450-500°). Based on this finding, it was possible to shorten the combustion tube by almost a half. By virtue of the decrease in the dead space of the apparatus gained through shortening the combustion tube, it has been possible to replace the usual 250-ml oxygen-rinse of the apparatus by a 10-minute rinse for a transmission rate of 10-15 ml oxygen per minute. The Mariotte bottle has been omitted, and, in the case of a low transmission rate, moderate suction with a water-jet vacuum pump has been introduced. The amount of oxygen pulled through by the water-jet pump is constant with time. In a rapid combustion, almost all of the oxygen is consumed to form water and CO_2, which are retained in the appropriate absorption tubes. If the amount of oxygen pulled through is to remain constant, the rate of oxygen flow through the catalyst must increase considerably as a result. The fact that correct results can be obtained even with this procedure is a further proof of the unusually high catalytic capacity of the decomposition product of $AgMnO_4$.

The absorptive capacity of the catalyst for chlorine and iodine was determined by capacity experiments. It is greatly superior to that of metallic silver, which is currently used for this purpose. It is advisable, however, to replace the combustion-tube filling after the analysis of only 5-10 S- or X-containing compounds, since the danger exists that the catalytic activity of the filling will be reduced as a result of absorption of large amounts of interfering elements.

Compounds containing fluorine give moderately high C-values and considerably high H-values. For this reason, the short filling in this form is not suitable for the analysis of F-containing compounds.

SUMMARY

The semimicrodetermination of carbon and hydrogen, using a 3.5-cm contact layer, is described.

LITERATURE CITED

[1] R. O. Clark and G. H. Stillson, *Anal. Chem.* **19**, 423 (1947).
[2] L. T. Hallett, *ibid.* **10**, 101 (1938).
[3] H. Reihlen, *Mikrochemie* **23**, 285 (1938).
[4] G. L. Royer, A. R. Norton, and O. E. Sundberg, *Ind. Eng. Chem., Anal. Ed.* **12**, 688 (1940).

[5] W. Zimmermann, *Mikrochemie ver. Mikrochem. Acta* **31**, 149 (1943).
[6] F. O. Fischer, *Anal. Chem.* **21**, 827 (1949).
[7] M. Večeřa and L. Synek, *Chem. Listy* **49**, 1891 (1955).
[8] J. Körbl, *Chem. Listy* **49**, 862 (1955); *Collection Czech. Chem. Commun.* **20**, 953 (1955).
[9] J. Körbl, *Chem. Listy* **49**, 858 (1955); *Collection Czech. Chem. Commun.* **20**, 948 (1955).
[10] F. Pregl: "Quantitative Organic Microanalysis", 4th English Ed., London (1945).
[11] V. Rýpar, Private communication.
[12] J. Körbl, *Chem. Listy* **49**, 929 (1955); *Collection Czech. Chem. Commun.* **20**, 993 (1955).

35

*The Thermal Decomposition Product of Silver Permanganate as a Combustion Catalyst in the Micro Carbon-Hydrogen Determination**

J. Körbl and K. Blabolil

With the increasing refinements in the procedures of organic chemistry, particularly in the field of natural products, it is becoming more and more necessary to carry out C and H determinations with minimum quantities of material. The first microdetermination of C and H was developed by Pregl [1], on the basis of the Liebig method. Over the years, a number of modifications have appeared, which have been critically evaluated in this country by Jureček. [2] The most recent knowledge in the field of microelemental analysis has been discussed by Kirsten. [3] Aside from the saving in material and energy, the advantages of microdeterminations over the analogous semimicro methods result mainly from the time factor. This is the main reason why micro methods are being introduced as routine determinations in all important laboratories, even in situations where semimicro methods can be used, although the micro methods require a more refined technique, better qualified workers, and better equipped laboratories.

In our preceding communication [4] we described the semimicrodetermination of C and H using a short catalytic filling of the thermal decomposition product

**Collection Czechoslov. Chem. Communs.*, 21, 318-20 (1956)

of $AgMnO_4$. [5] This method has now been used in the microdetermination. In view of the fact that the combustion time had already been reduced to 5-10 minutes in the semimicro method, transition to the micro scale does not lead to any significant additional saving of time. However, the method described retains all the advantages obtained by using the new contact material, as reported previously. [4, 6]

EXPERIMENTAL PART AND RESULTS

APPARATUS

The apparatus described in the preceding communication (Chap. 34) is used. [4] Between the $CaCl_2$ drying tower and the U-tube with bubble counter there is inserted a 10-mm-diameter glass tube, tapered at one end, and closed with a rubber stopper with inlet tube at the other. The tube is filled with a 3.5-cm layer of the decomposition product of $AgMnO_4$. The layer of contact material is held in place at both ends with ignited asbestos plugs. The tube serves for the combustion of any organic materials that may be present in the oxygen. [7] It is heated to the same temperature as the combustion-tube filling (450-500°).

Open Pregl tubes [1] were used for the absorption of water and CO_2.

N_2O_4 from N-containing compounds was retained in an adsorption tube, filled with MnO_2 adsorbed on silica gel, inserted between the H - and C- absorption tubes. [8] In the analysis of N-containing compounds, a heater [7] was used to heat the H-tube. [6]

A Mariotte bottle [1] was used to keep the quantities of oxygen passing through during the individual analyses constant.

Weighing was carried out in a thermostatted weighing room, on an undamped microbalance (Bunge, Hamburg).

ANALYTICAL PROCEDURE

About 4 mg of sample is weighed out. The oxygen stream is adjusted to 8 ml per minute and the compound is burned within about 5 minutes. After a brief ignition, the apparatus is rinsed with 80-100 ml oxygen. The total amount of oxygen passed through during an analysis, including rinsing, should be 150 ml.

The absorption tubes are weighed as follows: After disconnecting the apparatus, the tubes are wiped with a piece of moist flannel and then with two consecutive pieces of chamois skin. After cleaning the H-tube, the zero position of the balance is determined within 6 minutes. The H-tube is placed on the balance at the 6th minute, and the weight is read off at the 10th minute. The C-tube is placed on the balance at the 11th minute, and its weight is read off at the 15th minute. The same procedure is followed in the analysis.

Table 1 shows a series of consecutive analyses. The last four were carried out after an interruption of 5 weeks, and after refilling the H- and C- absorption tubes.

CONCLUSION

The catalytic oxidizing properties of the decomposition product of $AgMnO_4$ and its absorption properties for halogens and sulfur have been established in preceding papers [4-6] and the transition to the micro scale is not complicated by the combustion-tube filling.

TABLE 1

Compound	Formula	Calc. Found %C	%H	Δ %C	%H
Vanillin	$C_8H_8O_3$	63.15	5.30		
		63.34	5.52	+0.19	+0.22
Paraffin	$C_{29}H_{60}$	85.20	14.80		
		85.39	15.04	+0.19	+0 24
Cholesterol	$C_{27}H_{46}O$	83.86	11.99		
		84.02	12.15	+0.16	+0.16
Diphenyl-amine	$C_{12}H_{11}N$	85.17	6.55		
		85.16	6.81	-0.01	+0.26
Vanillin	$C_8H_8O_3$	63.15	5.30		
		63.34	5.53	+0.19	+0.23
Diphenyl-amine	$C_{12}H_{11}N$	85.17	6.55		
		85.17	6.79	0.00	+0.24
		85.17	6.78	0.00	+0.23
		85.28	6.82	+0.11	+0.27
Benzoic acid	$C_7H_6O_2$	68.85	4.95		
		68.92	5.11	+0.07	+0.16
Benzyl-thiuronium chloride	$C_8H_{11}ClN_2S$	47.40	5.47		
		47.46	5.66	+0.06	+0.19
		47.51	5.71	+0.11	+0.24
		47.59	5.64	+0.19	+0.17
Benzoic acid	$C_7H_6O_2$	68.85	4.95		
		69.01	5.14	+0.16	+0.19

Mean error of the determination: +0.11% C, +0.22% H
Mean deviation: 0.13% C, 0.22% H

The combustion of micro samples does not impose a serious burden on the catalytic filling, even for very rapid combustion. The small dimensions of the combustion tube and N_2O_4 absorption tube [8] permit work with small amounts of oxygen (a total of 150 ml). The complete determination takes 45 minutes.

Errors that had to be overcome in perfecting the method were mainly caused by the type of cleaning and weighing of the absorption tubes. By consistently following the procedure cited in the experimental part, correct results have been obtained. The mean error and mean deviation of 13 determinations shown in Table 1 are 0.11 and 0.13%, respectively, for carbon. The hydrogen results, on the average, are high by 0.22%. This constant error is not specific for the method described. High hydrogen values have been discussed by Lindner [7], among others. The error can be dealt with by applying a correction. Upon doing so with the analytical values for hydrogen shown in Table 1; i.e., correcting by -0.22%, the maximum error for hydrogen is 0.06%.

SUMMARY

The short catalytic combustion-tube filling, with some modifications, has been used in the micro carbon-hydrogen determination.

LITERATURE CITED

[1] F. Pregl: "Quantitative Organic Microanalysis," 4th English Ed., London (1945).
[2] M. Jureček, *Chem. Listy* **45**, 463 (1951).
[3] W. Kirsten, *Anal. Chem.* **25**, 74 (1953).

[4] J. Körbl, *Chem. Listy* **49**, 1532 (1955); *Collection Czech. Chem. Commun.* **20**, 1026 (1955).
[5] J. Körbl, Chem. Listy **49**, 858 (1955); *Collection Czech. Chem. Commun.* **20**, 948 (1955).
[6] J. Körbl, *Chem. Listy* **49**, 862 (1955); *Collection Czech. Chem. Commun.* **20**, 953 (1955).
[7] J. Lindner: "Mikromassanalytische Bestimmung des Kohlenstoffs und Wasserstoffs mit grundlegender Behandlung der Fehlerquellen in der Elementaranalyse," Berlin (1935).
[8] J. Körbl, Chem. Listy (in press).

36

The Complexometric Determination of Oxidizable Sulfur as Manganese(II) Sulfate

J. Körbl and R. Přibil

In a systematic study of the catalytic oxidizing properties of the thermal decomposition product of $AgMnO_4$ [1], one of us (Körbl) found that this material is not only an excellent oxidation catalyst, but also a very active absorbent for halogen and SO_2. Both of these properties have already been utilized in the determination of C and H in organic compounds. [2-4] Upon combustion in an oxygen stream, the SO_2 is retained quantitatively by the contact layer as Ag_2SO_4 and $MnSO_4$. [1]

It should be possible to determine the sulfates formed, after their extraction from the contact material, perhaps by their precipitation as $BaSO_4$. The only advantage of this method of determination over those generally used [5] would be a more rapid combustion. For this reason, a different approach has been considered. Since the amounts of Ag_2SO_4 and $MnSO_4$ formed correspond to the SO_2 absorbed, the problem of rapid S determination can be treated as a matter of the simultaneous determination of Ag and Mn(II) ions. The only volumetric method to be considered is complexometric titration. Mn can be determined complexometrically with great accuracy against Eriochrome Black T [6], but this does not

Collection Czechoslov. Chem. Communs.* **21, 322-26 (1956)

hold true for Ag. It was therefore necessary to find a suitable reaction which would make possible an exchange reaction of the Ag present in the solution after extraction of the contact mass against equivalent amounts of Mn. For this purpose we have used the reaction of Ag ions with MnO_2 and H_2O_2, which has hitherto not been used in analysis:

$$\text{(a)}\ 2Ag^{+} + MnO_2 + 2H_2O_2 = 2Ag + Mn^{++} + 2H_2O + 2O_2$$

The detailed study of the individual stages of this new and rapid S determination is the subject of this paper.

EXPERIMENTAL PART AND RESULTS

REAGENTS

The decomposition product of $AgMnO_4$ was prepared in the previously described manner. [1] For our experiments it was ignited for three hours at 500°.

MnO_2 was prepared by reaction of $Mn(NO_3)_2$ with $KMnO_4$ in alkaline solution. The precipitated MnO_2 was repeatedly washed by decantation with HNO_3 (1:10) and then with water until it gave no reaction for nitrate.

After filtration it was dried at 150°. The test for Mn (reaction with Eriochrome Black T in alkaline solution) with 1 g of the product extracted with 100 ml of distilled water was negative.

Urea peroxide hydrate [7] was prepared by crystallization of urea from 30% H_2O_2. The precipitated crystals were washed with a small amount of ice water, alcohol, and ether, and were dried in air. The solution of 1 g of this product in 10 ml of water was neutral to methyl orange.

The 0.05 N solution of disodium ethylenediaminetetraacetate was prepared in the usual manner from Chelaton 3 (VEB-Reagencie Company). A 0.05 N Ag_2SO_4 solution was prepared by dissolving the corresponding amount of Ag_2SO_4 in 1 liter of doubly distilled water. Its factor was determined complexometrically by the reaction with $K_2Ni(CN)_4$ according to Flaschka. [8]

Eriochrome Black T was used in the solid state, mixed with NaCl in the ratio 1: 1000.

The buffer solution was prepared by dissolving 54 g NH_4Cl in 350 ml concentrated NH_4OH and making up to 1 liter with twice distilled water.

All reagents used were analytically pure.

The apparatus used was the same as in the carbon-hydrogen determination by the "short combustion-tube filling" method [3], but without absorption tubes.

COMPLEXOMETRIC DETERMINATION OF SILVER

It was proved by numerous experiments that reaction (a) proceeds unambiguously only in neutral solution. In acid solution it is accompanied by the known reaction:

$$\text{(b)} \quad MnO_2 + 2H^+ + H_2O_2 = Mn^{++} + 2H_2O + O_2$$

For this reason the use of commercial H_2O_2 preparations, which are always weakly acid, gave high Mn values that did not agree with the amounts of Ag_2SO_4 used in the reaction. Solid urea peroxide hydrate proved much better than neutralized H_2O_2, and was used in all subsequent experiments.

It was also shown, by a series of control experiments, that reaction (b) already occurs to a measurable extent in the presence of CO_2. Twice distilled water, freed of CO_2 by boiling, was therefore used in subsequent experiments.

The smooth course of reaction (a) requires complete contact of the solid phase (MnO_2) with the solution. It has been shown by experiment that a quantitative reaction is obtained by treating the solution, with intensive stirring, at least three consecutive times alternately with solid MnO_2 and with urea peroxide hydrate. Twice the theoretical amount of MnO_2 and about 0.5 g of urea peroxide hydrate are sufficient for conversion. The reagents, if in excess, have no effect on the results. The course of the reaction is easy to follow visually. At first, MnO_2 coated with precipitated Ag rapidly sinks to the bottom in large flakes, and the supernatant liquid is clear. After the reaction is complete, the finely divided MnO_2 floats in the solution.

After final treatment, the complexometric Ag determination is carried out as follows: The pipetted 0.05 N Ag_2SO_4 (10 or 25 ml) is diluted to 50 ml and portions under constant stirring are alternately treated with MnO_2 and urea peroxide hydrate (0.5 g) several times in succession. After the spontaneous decomposition of the H_2O_2, the solution is filtered (Weissband filter) and the residue on the filter is washed with 200 ml of cold water, from which CO_2 has been boiled out. To the filtrate are added a few crystals of $NH_2OH \cdot HCl$, 2-3 ml of buffer solution, and the indicator, and the Mn is titrated with 0.05 N Complexon solution. When the listed conditions are observed, filtration is rapid, the filtrate is clear, and no opalescence appears (traces of Ag) after the addition of $NH_2OH \cdot HCl$. The results obtained in the manner described are shown in Table 1.

TABLE 1

Complexometric determination of silver

ml 0.05 N Ag_2SO_4 taken	ml 0.05 N Complexon used: Flaschka method [8]	ml 0.05 N Complexon used: Mn^{++} titration
10	4.75	4.78; 4.75; 4.80; 4.74
25	11.88; 11.90	11.93; 11.90; 11.80; 11.90

DETERMINATION OF OXIDIZABLE SULFUR

As discussed above, the burned sulfur is absorbed quantitatively in the contact material in the form of Ag_2SO_4 and $MnSO_4$, which can be extracted with water. However, the decomposition product of $AgMnO_4$ used must not contain any Ag_2O, which is also soluble in water. (The decomposition product obtained by gently heating $AgMnO_4$ in a test tube [1] still contains considerable quantities of free Ag_2O, which can be decomposed only by prolonged ignition at 500°.)

Ignition in the combustion tube during the determination proper is not sufficient for complete decomposition. For this reason, a product ignited for three hours at 500° is used for the actual tube filling. The preparation obtained in this manner meets the requirements stated above. Yet its chemical behavior, contrary to expectation, does not correspond to a mixture of MnO_2 and elemental Ag, in spite of the fact that its composition corresponds to the formula $AgMnO_2$. [1] By the action of dilute acids—including CO_2 —both Ag and Mn are liberated. The reaction can be expressed by the following equation:

(c) $2AgMnO_2 + 4H^+ = 2Ag^+ + MnO_2 + Mn^{++} + 2H_2O$

This reaction pattern was confirmed by exhaustive extraction of the decomposition product with dilute HNO_3. It is therefore necessary to exclude the action of CO_2 also while washing the decomposition product.

Check analyses for sulfur were carried out with various samples of pure sulfur melt. The procedure was as follows: A small, ignited asbestos plug was placed in the tip of the combustion tube, and the tube was filled with a 3.5-cm layer of the decomposition product, ignited at 500°. The filling was heated to 450-500° with a microburner, and the oxygen stream was set at 20 ml per minute. The platinum boat with the weighed sulfur sample was then placed in the tube, the sample was burned within 1-2 minutes, and the tube was briefly ignited. After cooling, the filling was poured into a beaker, the tube was rinsed with 50 ml of hot water, and the solution was briefly boiled. After this had cooled, the reaction with MnO_2* and urea peroxide hydrate was carried out, and the Mn was determined complexometrically, as described in the preceding section. The blank was carried out in the same manner. Some of the results obtained are shown in Table 2.

CONCLUSION

The complexometric Ag determination based on the exchange reaction of Ag ions with Mn(II) ions is, in itself, only of theoretical interest. However, when combined with the absorption of SO_2 in the decomposition product of $AgMnO_4$, it represents a completely

* The MnO_2 contained in the excess decomposition product is not always sufficient for complete conversion. It is therefore necessary to proceed as in the determination of pure silver solutions.

TABLE 2

Determination of Oxidizable Sulfur

mg S taken	ml 0.05 N Complexon used*	mg S found	Difference (mg)
22.15	13.85	22.23	0.08
19.65	12.30	19.74	0.09
12.22	7.65	12.27	0.05
4.08	2.55	4.09	0.01

* Factor 1.0012; the blank (0.10 ml) was subtracted from the amount used.

new prin ple for determining oxidizable sulfur. The main advantage of the new method described is its speed, in which it exceeds all methods based on the gravimetric or titrimetric [9] determination of precipitated $BaSO_4$. The advantage of the new method over methods for absorbing SO_2 on metallic silver [10-14] lies in the far higher absorptivity of the decomposition product, which makes it possible to burn the analytical sample in a considerably shorter time. The sulfur samples were burned within 1-2 minutes at an oxygen flow rate of 20 ml per minute. Furthermore, the complexometric determination of Mn is much simpler and more rapid than the direct determination of Ag. The whole determination can be performed comfortably in 30 minutes.

It is to be expected that the determination will also be applicable to special problems, such as the determination of volatile sulfur compounds in air or in gaseous reaction products (determination of CS_2, SO_2, mustard gas, etc.)

The determination of sulfur in organic compounds, pyrites, etc., is currently being studied in detail.

SUMMARY

A complexometric method for determining oxidizable sulfur after absorption of SO_2 in the thermal decomposition product of $AgMnO_4$ has been developed. SO_2 is absorbed quantitatively in the form of Ag_2SO_4 and $MnSO_4$. By reaction with H_2O_2 and MnO_2, the Ag_2SO_4 is converted to an equivalent amount of $MnSO_4$, which is then determined complexometrically.

LITERATURE CITED

[1] J. Körbl, *Chem. Listy* **49**, 858 (1955); *Collection Czech. Chem. Commun.* **20**, 948 (1955).

[2] J. Körbl, *Chem. Listy*. **49**, 862 (1955); *Collection Czech. Chem. Commun.* **20**, 953 (1955).

[3] J. Körbl, *Chem. Listy* **49**, 1532 (1955); *Collection Czech. Chem. Commun.* **20**, 1026 (1955).

[4] J. Körbl and K. Blabolil, *Chem. Listy* **49**, 1664 (1955); *Collection Czech. Chem. Commun.*, **21** 318 (1956).

[5] M. Jureček: "Organic Analysis," Prague (1950) p. 169.

[6] H. Flaschka and A. M. Amin, *Mikrochim. Acta* **1953**, 414.

[7] S. Tanatar, *Zh. Russ. Fiz.-Khim. Obshchestva*, 40 376 (1908).

[8] H. Flaschka and F. Huditz. *Z. Anal. Chem.*. **137**, 104 (1952).

[9] R. Belcher, D. Gibbons, and T. S West, *Chem. and Ind.* (*London*) 1954, 127.

[10] E. W. D. Huffman, *Ind. Eng. Chem., Anal. Ed.* **12**, 53 (1940).

[11] R. Belcher and C E Spooner, *J. Chem. Soc.* 1943, 313.

[12] G. L. Stragand and H W. Safford, *Anal. Chem.* **21**, 625 (1949).
[13] F. Zinneke, *Z. Anal. Chem.* **132**, 175 (1951).
[14] M. Večeřa, *Chem. Listy* **48**, 613 (1954).

37

*A Simple and Rapid Method of Qualitative Elemental Organic Analysis**

J. Körbl and R. Přibil

The need for a special test for carbon in qualitative organic analysis arises only for a limited number of compounds. [1, 2] The detection is usually accomplished by igniting the material with CuO and reacting the CO_2 formed with $Ba(OH)_2$. [2, 3] Emich [4] decomposed the material by heating in a sealed capillary, where the carbon is usually detected by the formation of a bright mirror. Several other detection methods have also been proposed. [1] Of much greater importance for analytical practice is, however, the detection of other elements, particularly nitrogen, sulfur, and halogens. The best known nitrogen test is the Lassaigne reaction. [5] The advantage of the Lassaigne test is that a number of other elements can be detected in the fusion mixture. [6] A disadvantage is the use of an alkali metal: although the danger of explosion in the destruction of the excess metal (particularly potassium) can be considerably reduced by working on a micro scale [6, 7], the technique is still involved. Nitrogen can also be detected through NH_3 formation upon heating the compound with CaO. Emich [4] used a mixture of CaO and zinc, and also detected

**Collection Czechoslov. Chem. Communs.* 21, 955-59 (1956)

sulfur and halogens after extraction. On this basis, Bennett [8] developed a systematic micro method, which includes tests for nitrogen, chlorine, bromine, iodine, sulfur, arsenic, and phosphorus. However, Widmark [9] found that upon heating the compound with CaO and zinc, a small amount of cyanide is formed along with the NH_3, which interferes with the detection of halogen. Nitrogen and a number of other elements can be detected by fusion with a mixture of magnesium metal and K_2CO_3 [10, 11], and possibly also saccharose. [12] However, these methods do not prevent the reaction of metallic magnesium with atmospheric nitrogen. [13] The best known halogen test, aside from those already mentioned, is the Beilstein test, but this also gives a positive result for several compounds containing nitrogen and oxygen. [1-3] Much more reliable, for example, is the reaction as modified by Hayman. [14] However, this cannot provide information as to whether halogen is present only in traces, as an impurity, or in larger amounts.

In the first communication in this series [15], we drew attention to the reactions of organic compounds with the decomposition product of $AgMnO_4$. The catalytic oxidative properties of the decomposition product have already been used in a new C and H determination [16-18], and the absorptive properties for absorbing N_2O_4 [16] in the C and H determination and also for the complexometric determination of oxidizable sulfur. [19]

In the present work we have been concerned with using both the degradative and absorptive properties of the decomposition product of $AgMnO_4$ for the qualitative detection of elements in organic compounds. The proposed method is based on the following facts: Heating an organic compound with an excess of the decomposition product of $AgMnO_4$ results in an oxidative degradation of the material, already at a moderately high temperature. This is usually observable in the pyrophoric behavior of the reaction mass or of some of its particles. Carbon and hydrogen are burned to CO_2 and H_2O; mercury is liberated as the element. Sulfur is retained in the reaction mixture as Ag_2SO_4 and $MnSO_4$, halogens as silver halides, and part of the nitrogen as $AgNO_3$. Phosphorus and arsenic are retained in the form of phosphate and arsenate. It follows from this that many of the frequently sought elements are present in the form of silver salts, which differ in their solubilities in water, ammonia, and the like. These can therefore be extracted consecutively from the reaction mixture in a suitable manner, and can be detected by sensitive reactions of qualitative analysis.

The execution of these reactions and the experimental technique that makes possible convenient and reliable work on a micro scale are described in detail in the experimental part.

EXPERIMENTAL PART AND RESULTS

APPARATUS AND REAGENTS

Decomposition and filtration tube (Fig. 1): A 13-cm glass tube of 4 mm i. d. is constricted to about 0.3 mm at about 9 cm from the upper end, and the shorter part is drawn out to a pipette. Glass tubes of the same diameter, drawn out to a long capillary, serve as micropipettes. Liquids are handled by means of 15-cm capillaries, 0. 5 mm in diameter. For the determination of arsenic, the apparatus of Văsák and Šedivec is used. [20]

The decomposition product of $AgMnO_4$ is prepared by heating $AgMnO_4$ in a test tube. [15]

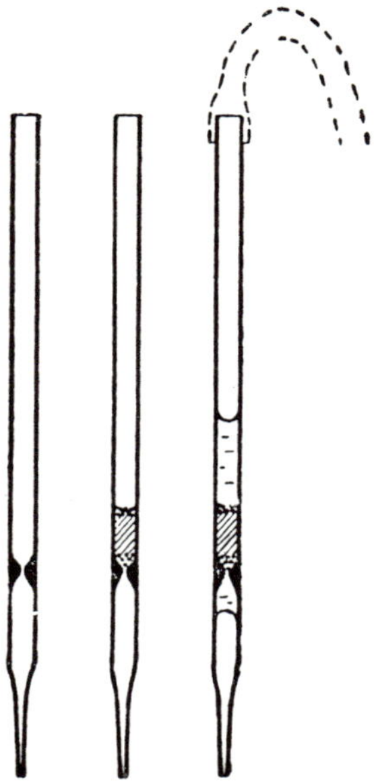

Fig. 1 Decomposition and Filtration Tube.
a: Empty; b: filled; c: during washing out.

PROCEDURE

Sample Preparation

0.1-1.0 mg of the compound under examination is mixed with about 50 mg of the decomposition product of $AgMnO_4$ on a mixing paper, and is ground with a metal spatula. A filter charge made of a small, ignited asbestos plug (fibrous asbestos for Gooch crucibles) is placed in the decomposition tube, and is moderately compacted in the constricted part with the handle of the metal spatula. The prepared sample is poured into the tube and tapped down. The approximately 0.5-1-cm layer formed is again secured with an asbestos plug (Fig. 1b). If the compound under investigation is liquid, a microdrop is sucked into the capillary, the outside of the capillary is wiped with filter paper, the contents are moved on to the layer of decomposition product with the filter charge, covered with a small additional amount of decomposition product, and secured with asbestos. After moderate heating with a microburner, decomposition is complete within a few seconds. The individual elements are detected as follows:

Hydrogen

Hydrogen is identified by the formation of dew on the cool wall of the decomposition tube.

Carbon

After the reaction, a quantity of CO_2 sufficient for reliable identification remains in the decomposition tube. A rubber bulb (e.g., from an automatic burette or the like) is attached to the upper end of the still warm tube, and the gaseous contents of the tube are forced at moderate speed directly over the surface of a clear, freshly filtered, saturated $Ba(OH)_2$ solution (about 1 ml) in a 12 x 120 mm test tube. The CO_2 is detected by a turbidity of $BaCO_3$, readily visible on the liquid surface. In order to prevent the CO_2 from "pouring out", the decomposition tube must be kept in a vertical position during the time between heating the sample and testing for CO_2.

Nitrogen

The N_2O_4 formed in varying quantities during the oxidation of nitrogen-containing compounds is partly retained in the decomposition product and partly dissolved in the droplets of water on the wall of the tube.

A few drops of distilled water are introduced into the cooled tube with a micropipette, and the liquid is forced into the tapered end by means of the rubber bulb (Fig. 1c). One drop of liquid is added to 1 ml of a 1% solution of diphenylamine in concentrated H_2SO_4 on a watch glass. The presence of nitrogen is shown by a blue coloration at the surface of contact, which is particularly easy to see against a white background.

Sulfur

The remainder of the liquid is forced out of the tapered end of the tube into a micro test tube. The presence of sulfur is shown by the formation of a white precipitate of $BaSO_4$ when a drop of 5% $Ba(NO_3)_2$ solution is added.

Chlorine and Bromine

After the tests described above, the tube is rinsed out with water and charged with a few drops of NH_4OH (1:1); the solution is then forced into a test tube. After acidification with concentrated HNO_3, a characteristic turbidity or a precipitate of AgCl or AgBr is formed. The halogen is identified by one of the standard reactions, e.g., by means of the micro method of Wilson and Wilson. [21]

Phosphorus and Arsenic

Phosphorus or arsenic is detected with molybdate solution in the filtrate or centrifugate after the silver halide precipitate. Arsenic can also be detected by the method of Văsák and Šedivec [20], or directly in a drop of the ammoniacal filtrate.

Iodine

After washing with NH_4OH, which, if the test for chlorine or bromine is positive, is repeated until a negative reaction is obtained, the residue in the tube is washed out with a few drops of 30% $AgNO_3$ solution, in which AgI is readily soluble. Upon dilution of the filtrate with water, the solution first becomes turbid from the precipitated silver iodide-nitrate complex; AgI then precipitates on further dilution.

Mercury

If the sample is believed to contain mercury, the decomposition procedure must be slightly modified. The tube is filled in the manner described, and a moderate current of air is simultaneously passed through during heating by means of the rubber bulb. A small amount of mercury, sufficient for detection, volatilizes into the tapered part of the tube, where it is visible to the naked eye. A reliable identification is carried out by sucking a small amount of dithizone solution (1.3 mg in 100 ml CCl_4) into the tip of the decomposition tube. The solution sucked in must not come in contact with the asbestos filling. In the presence of mercury, a color change from green to the characteristic orange-yellow is observed. Since a reaction of metallic mercury with dithizone is involved, the color change is somewhat slower (within about one minute). By blowing out and repeated sucking in of a little water, the dithizone solution is removed from the tip of the tube, again taking care to prevent contact of the liquid with the filter charge. The other tests, except for carbon, are carried out as described above. If a test for carbon must also be made, the tip of the tube is connected by thin rubber tubing to a suitably bent capillary, whose end dips into a test tube filled with $Ba(OH)_2$. The decomposition is then carried out while blowing air through the tube. After decomposition is complete, the capillary is removed and mercury is detected.

DISCUSSION

The proposed scheme of qualitative analysis makes possible a rapid and convenient determination of the elemental composition of organic compounds. In its total execution it is particularly suitable for training and educational purposes, since the need for determining the presence of all possible elements arises only very rarely in actual practice. Although the method is described as a "systematic" detection of elements, the performance of any given individual test is practically independent of the presence of other elements. To be sure, the most frequent task in analytical practice is, as a rule, the identification of one or at most two elements, even when an "unknown" compound, e.g., a by-product of a synthesis, is involved.

In view of the fact that the reaction mass is in each case washed out with only a few drops of the corresponding reagent, and that one therefore works with relatively concentrated solutions, the reactions used achieve their maximum sensitivities. If the element tested for is present as a constituent of the compound under investigation and not as an impurity, 1 mg of material is sufficient for a reliable detection in the usual molecular weight range.

The method described does not, of course, represent a closed operation. If necessary, the scheme of analysis can be extended to include a few additional elements. On the basis of previous experience, we assume that it will be possible to modify this method without any serious difficulty for the quantitative determination of several elements. We are continuing work along this line.

SUMMARY

A simple micro method has been developed for the detection of carbon, hydrogen, mercury, nitrogen, chlorine, bromine, iodine, sulfur, phosphorus, and arsenic in organic materials. A 0.1-1-mg sample, mixed with the decomposition product of $AgMnO_4$, is oxidized by gentle warming. Hydrogen is detected by the formation of dew in the reaction tube. Carbon is detected by reaction with barium hydroxide. Mercury forms a coating on the tube wall, in which it is detected with dithizone. Halogens, sulfur, phosphorus, arsenic, and part of the nitrogen are retained in the reaction mass mainly as their silver salts and are brought into solution by stepwise extraction with water, ammonia, and silver nitrate, before identification. The decomposition of the material and the extraction take place in a tube adapted for this purpose, so that filtration is unnecessary. All detections can be performed within a few minutes.

LITERATURE CITED

[1] F. Schneider: "Qualitative Organic Microanalysis", New York (1946).
[2] N. D. Cheronis: "Micro and Semimicro Methods", New York (1954).
[3] M. Jureček: "Organic Analysis", Prague (1950).
[4] F. Emich and F. Schneider: "Microchemical Laboratory Manual", New York (1932).

[5] J. L. Lassaigne, *Compt. rend.* **16**, 387 (1843).
[6] G. Widmark, *Acta Chem. Scand.* **7**, 1395 (1953).
[7] F. Feigl: "Spot Tests", Vol. II, 4th Ed., New York (1954).
[8] E. L. Bennett, et al., *Anal. Chem.* **19**, 1035 (1947).
[9] G. Widmark, *Acta Chem. Scand.* **8**, 246 (1954).
[10] R. H. Baker and C. Barkenbus, *Ind. Eng. Chem., Anal. Ed.* **9** 135 (1937).
[11] D. G. Foulke and F. Schneider, *ibid.* **10**, 104 (1938).
[12] V. Castellana, *Gazz. Chim. Ital.* **34 II**, 357 (1904).
[13] E. H. Russel, *Chem. News* **102**, 187 (1910).
[14] D. F. Hayman, *Ind. Eng. Chem., Anal. Ed.* **11**, 470 (1939).
[15] J. Körbl, *Chem. Listy* **49**, 858 (1955); *Collection Czech. Chem. Commun.* **20**, 948 (1955).
[16] J. Körbl, *Chem. Listy* **49**, 862 (1955); *Collection Czech. Chem. Commun.* **20**, 953 (1955).
[17] J. Körbl, Chem. Listy **49**, 1532 (1955); *Collection Czech. Chem. Commun.* **20**, 1026 (1955).
[18] J. Körbl and K. Blabolil, *Chem. Listy* **49**, 1664 (1955); *Collection Czech. Chem. Commun.* **21**, 318 (1956).
[19] J. Körbl and R. Přibil, *Chem. Listy* **50**, 232 (1956); *Collection Czech. Chem. Commun.* **21**, 322 (1956).
[20] V. Vašák and V. Šedivec, *Chem. Listy* **46**, 341 (1952).
[21] D. W. Wilson and C. L. Wilson, *J. Chem. Soc.* **1939**, 1956.

38

An Investigation into the Nature of the Thermal Decomposition of Silver Permanganate *

V. Šatava and J. Körbl

The thermal decomposition of silver permanganate has already been investigated previously by a number of authors [1-7], from the standpoint of both reaction kinetics [2, 4, 6, 7] and composition of the material formed by the decomposition [1-3, 5]. The decomposition product of silver permanganate aroused our interest in connection with the discovery of its catalytic oxidative and adsorbent properties. [8] As a first step towards a more detailed knowledge of catalysis mechanisms, we have followed the further transformations of the decomposition product up to high temperatures; we considered this study necessary since the operating temperatures for the potential applications described lie between 450 and 600°C. [9-13] Earlier papers [1-7] dealt only with the first stage of decomposition, which proceeds autocatalytically [6, 7] at 100°C, or with the product of the spontaneous glow decomposition [3], which occurs on rapid heating of silver permanganate to a somewhat higher temperature.

The black, water-insoluble, amorphous [3] products of the isothermal (100°C) and the spontaneous

**Collection Czechoslov. Chem. Communs.* 22, 1380-89 (1957)

decompositions are only slightly different; in both, the overall composition approximates the formula [3] $AgMnO_{2.66}$. Upon solution in a mixture of dilute H_2SO_4 and oxalic acid or $FeSO_4$, part of the oxygen escapes as gas (inactive oxygen). Gorgeu attributes this phenomenon to the presence of silver(II) oxide, AgO[1], which is said to behave similarly. However, the action of concentrated HNO_3 on the decomposition product does not produce the brown coloration characteristic for Ag(II) salts. [2] Hein [3] assumes a common structural basis with the formula

$$8\ Ag_2O \cdot 15\ MnO_2 \cdot MnO_3$$

for decomposition products of silver permanganate prepared in different ways; it is modified by the presence of water or different amounts of inactive oxygen, about whose type of bonding he gives no further information. Prout and Tompkins [4] are of the opinion that the decomposition of silver permanganate can best be represented by the equation:

$$2\ AgMnO_4 = Ag_2O + 2\ MnO_2 + 1.5\ O_2$$

Grant and Katz [5] assume a mixture of the compound $AgMnO_3$, metallic silver, Ag_2O, and MnO_2 in the decomposition product.

As seen from these statements, previous ideas concerning the composition of the decomposition product of silver permanganate are quite varied. They are based on more or less accurate chemical analyses, for whose interpretation the classical concepts of inorganic chemistry were used. These notions could not lead to a complete understanding of the composition and structure of an amorphous material such as the decomposition product of silver permanganate.

We have therefore followed the decomposition and the other changes appearing at higher temperatures, which had already been observed in part previously [8], by the methods of thermogravimetric and differential thermal analysis, by measurements of electrical conductivity, by dilatometry, and by other methods.

EXPERIMENTAL PART AND RESULTS

Silver permanganate was prepared according to a previously described procedure [8], but was dried only over P_2O_5 in vacuum and stored in the dark. The isothermal decomposition product of silver permanganate was prepared by heating thin silver permanganate layers to various temperatures for 24 hours in a drying oven. Individual samples had the composition $AgMnO_{2.76-2.60}$. The glow decomposition product of silver permanganate was prepared by heating to the temperature of the spontaneous decomposition and had the composition $AgMnO_{2.58}$. The other chemicals used were of analytical grade.

The results of our preliminary experiments can be summarized as follows: Individual samples of both the isothermal decomposition product and the product formed by glow decomposition give a faint green coloration of variable intensity with concentrated alkali solution, which is a proof of the presence of traces of Mn_2O_3. For some samples obtained by glow decomposition, this reaction was negative. From the low-temperature decomposition products, small quantities of Ag_2O are gradually extracted with boiling water. With alkali chloride solution, the products react to form AgCl and liberate alkali hydroxides; the reaction with $MgCl_2$ solution is similar. With KI solution, the reaction leads to complete decomposition, forming KOH, iodine, AgI, and manganese iodide. Analogous but considerably slower reactions take place with the product ignited to 500°C.

From the decomposition products prepared at low temperature, one-third to four-fifths of the amount of silver present can be extracted with acids (dilute or concentrated CH_3COOH or HNO_3), depending upon the acid concentration, the reaction time, temperature, degree of subdivision, etc. Even after long contact with HNO_3, it was not possible to extract the final residue (1%) of silver. In addition to the silver, a very small amount of manganese is also extracted, in contrast to the decomposition product of $KMnO_4$, where no extraction of manganese takes place under similar conditions. In the products ignited to 500°C, half of the total manganese goes into solution, along with the silver. Inactive oxygen [1] was detected in both the isothermal (100°C) and glow decomposition products of either $AgMnO_4$ or other permanganates, e.g., $KMnO_4$ and $LiMnO_4$. After heating for several hours to only 110°C, the inactive oxygen had vanished except for an insignificant residue (of the order of 0.1%).

METHOD

Differential Thermal Analysis (DTA)

This was carried out in the usual manner. [14] The rate of temperature increase was 10°C/min, the temperature and temperature differences were measured with Pt-PtRh thermocouples, and the curves were recorded with a LeChatelier-Saladin apparatus. Six-tenths of a gram samples were used, and crystalline corundum was used as a standard. For analysis in a hydrogen stream, the comparison substance and sample were enclosed in a porcelain tube, through which was passed a constant hydrogen stream.

Thermogravimetric Analysis (TGA)

The curves were obtained with a photographic recorder [15], the temperature was measured directly in the sample with a Pt-PtRh thermocouple, and the temperature rise (6°C/min) was controlled automatically.

Contraction Thermal Analysis

A rod of dimensions 50 × 5 × 2 mm was pressed from the decomposition product at a pressure of 1 ton/sq cm. The changes in length with increasing temperature were measured in a simple dilatometer. The rate of temperature rise was 6°C/min.

Electrical Conductivity

For measuring changes of conductivity with temperature, a tablet was pressed at a pressure of 10 tons/sq cm. Silver-wire electrodes were inserted into two accurately drilled holes. The tablet was placed into a corundum block (to ensure uniform heat distribution) and the block was placed in an electric furnace. A linear temperature rise of 10°C/min was controlled automatically. The resistance of the sample was measured by a compensation method.

Thermoelectric Behavior

The thermoelectric behavior of the decomposition products was determined on the same tablets as the electrical conductivity, by measuring the EMF upon heating one of the two silver electrodes near the tablet surface. [16] Since the resistance of the sample was high, a vacuum-tube voltmeter had to be used for the measurement.

X-ray Structure Determination

This was carried out with FeK_α rays (Mn filter) in a 115-mm diameter Debye-Scherrer chamber, using a 12-hour exposure at 40 kV and 20 mA.

Specific Surface Area

The specific surface area of the product ignited to 500°C was measured by the Brunauer-Emmett-Teller method [17], using butane.

RESULTS

The differential thermal analysis of silver permanganate is shown in curve 1 of Fig. 1. The maximum at 160°C corresponds to the first decomposition. Further transformations are shown in the curve by endothermal minima at 300, 480, 630, 780, 940, and 960°C. In the first decomposition, partial dispersion of the sample occurs, and the rapid heat evolution causes overheating. For this reason, thermal analyses were also performed for the products obtained by glow decomposition (curve 2) and by isothermal decomposition at 100°C (curve 3).

The products corresponding to temperatures of 100, 500, and 650°C were investigated by an X-ray method. All of them were X-ray-amorphous even after heating for 6 hours at the temperature in question. In order to explain the significance of the individual minima on the curves, the thermal analysis of pure Ag_2O was therefore carried out (curve 4). This shows a decomposition minimum at 450°C, and another minimum, corresponding to the melting of metallic silver, at 960°C. Curve 5 was obtained with AgO; this shows a new minimum at 480°C, when compared with Ag_2O. Curve 6 represents MnO_2; the first minimum at 530°C denotes transition to Mn_2O_3, another minimum occurs at 850°, and, finally, transition to Mn_3O_4 is found at 960°C.

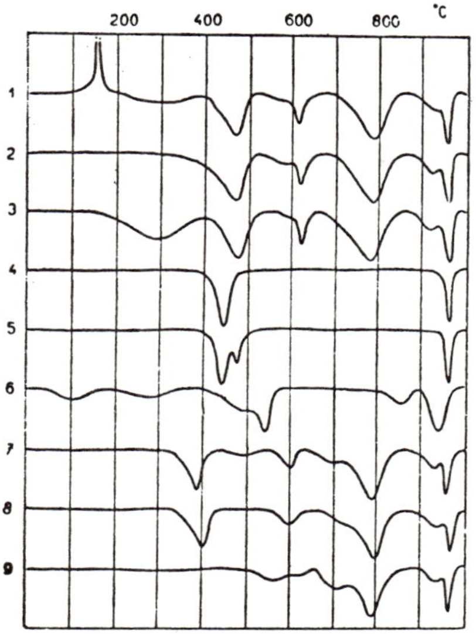

Fig. 1 Differential thermal analyses.
1: $AgMnO_4$; 2: product of glow decomposition; 3: product of isothermal decomposition at 100°C; 4: Ag_2O; 5: AgO; 6: MnO_2; 7: mixture of Ag_2O and MnO_2; 8: mixture of AgO and MnO_2; 9: mixture of metallic silver and MnO_2.

From these curves already, the significance of some extremes on the DTA curve of silver permanganate can be determined. At 960°C, fusion of the silver liberated in the decomposition takes place. The minimum at 940°C indicates the transition of Mn_2O_3 to Mn_3O_4. We would be most inclined to attribute the reaction at 480°C to decomposition of the Ag_2O present, but the temperature difference is large (30°C). Curve 7, which was obtained by analysis of a finely ground mixture of Ag_2O and MnO_2 in the mole ratio 1:2, shows that MnO_2, as is well known, catalyzes the decomposition of Ag_2O, which then takes place at only 380°C.

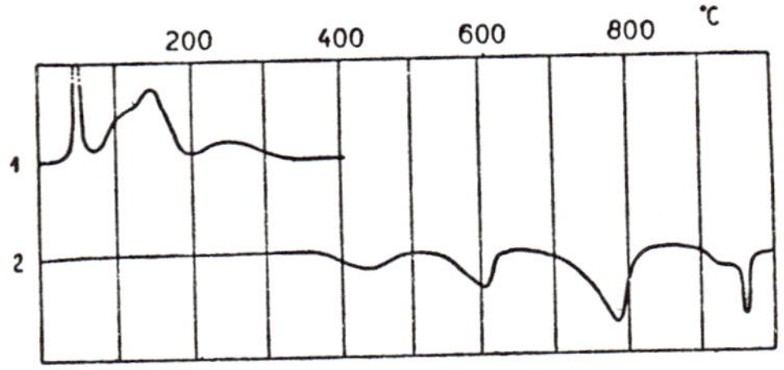

Fig. 2 Differential thermal analysis. 1: Isothermal decomposition product (100°C); analysis in a hydrogen stream; 2: product reduced with hydrogen to the first maximum in curve 1; analysis in air.

Likewise, curve 8, obtained by analysis of a mixture of AgO and MnO_2, precludes the possibility of the presence of AgO as an independent phase in the decomposition product of $AgMnO_4$.

Further information on the minimum at 480°C is given by the thermal analysis of $AgMnO_4$ in a hydrogen stream, as shown in Fig. 2. The first sharp maximum at 60°C indicates reduction of all the silver present to the metal, as proved by measuring the hydrogen consumption. The product of the reaction carried out up to the first maximum is silvery; its DTA curve (in air) no longer shows a minimum at 480°C (Fig. 2).

Similarly, this minimum disappears almost completely on extraction with a $MgCl_2$ solution, and is replaced by a sharp maximum at 455°C, corresponding to melting of the AgCl formed. This therefore involves a loss of oxygen, which, however, is bound to the silver more strongly than in Ag_2O.

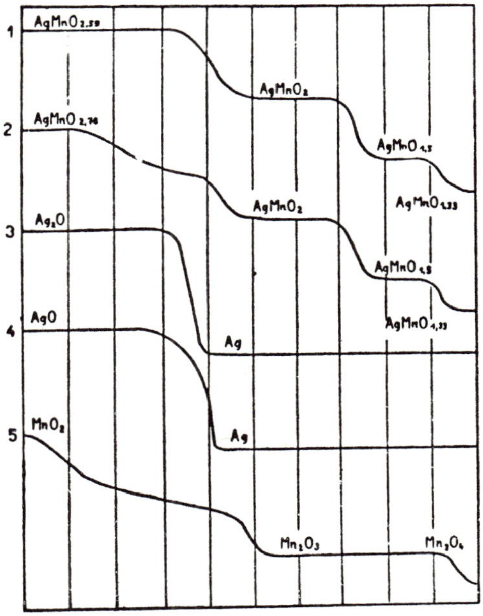

Fig. 3 Thermogravimetric analyses. 1: Product of glow decomposition; 2: product of isothermal decomposition; 3: Ag_2O; 4: AgO; 5: MnO_2.

The last curve (9) in Fig. 1 shows the thermal analysis of a mixture of metallic silver and MnO_2. From curves 7, 8 and 9 of Fig. 1 it can be seen that the presence of silver prevents the decomposition of MnO_2 to Mn_2O_3, which should take place at 530°C. Only the last curve shows an indication of this decomposition, which is probably a result of the coarse dispersion of both materials.

A further step towards the explanation of the processes occurring in the decomposition of $AgMnO_4$ consisted in carrying out thermogravimetric analyses, which are shown in Fig. 3. Curve 1 belongs to the decomposition product obtained by glow decomposition, curve 2 to that obtained by isothermal decomposition at 100°C. The thermogravimetric analysis of $AgMnO_4$ could not be carried out directly, since the first decomposition is too vigorous and part of the sample is dispersed even when it is covered with an inert material. Curve 3 was obtained with Ag_2O, curve 4 with AgO, and curve 5 with MnO_2.

The course of oxygen evolution after the first exothermal decomposition depends to a certain extent on the experimental conditions, as shown by the DTA and TGA curves of the products obtained by glow decomposition, by decomposition in an autoclave, and by isothermal decomposition of $AgMnO_4$ at 100, 200, and 300°C (Fig. 4). It is seen from the curves that the decomposition conditions affect only the first oxygen loss. The oxygen loss at 480°C is not significantly changed. The overall composition of the decomposition products at the individual temperatures is shown on the curves.

The thermogravimetric analyses confirm the significance of the minimum at 950°C in the DTA curve of $AgMnO_4$ (Fig. 1), and show that the decomposition to silver and Mn_2O_3 starts only at 780°C.

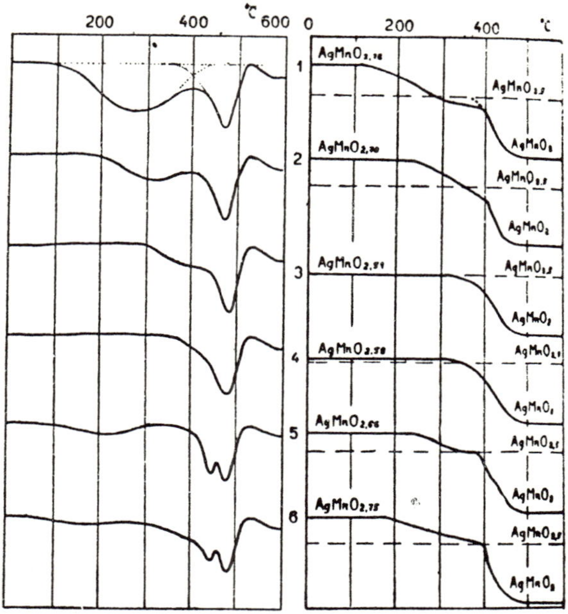

Fig. 4 Differential thermal and thermogravimetric analyses of decomposition products of $AgMnO_4$ prepared in different ways. 1: Isothermal decompositon at 100°C; 2: 200°C; 3: 300°C; 4: product of glow decomposition; 5: product of decomposition in autoclave at 150°C; 6: product of decomposition in autoclave at 110°C.

The minimum at 480°C is related to an oxygen loss, and the final product has the composition of $AgMnO_2$. The chemical reactions that were to determine whether the product is a mixture of metallic silver with MnO_2, or the compound $AgMnO_2$, gave no results, since MnO_2, as shown in a control experiment, oxidizes silver in chemical reactions, when mixed with it. The answer was obtained by measuring the temperature dependence of the electrical conductivity. The curve of 1/R vs. 1/T obtained is shown for the isothermal product in Fig. 5. The increase in electrical conductivity up to a temperature of 415°C shows the semiconductor character of the decomposition product. Thermoelectric measurement proved that the conductivity is of the P type, and that the electricity is therefore conducted by positively charged lattice holes of the solid material. The sharp break

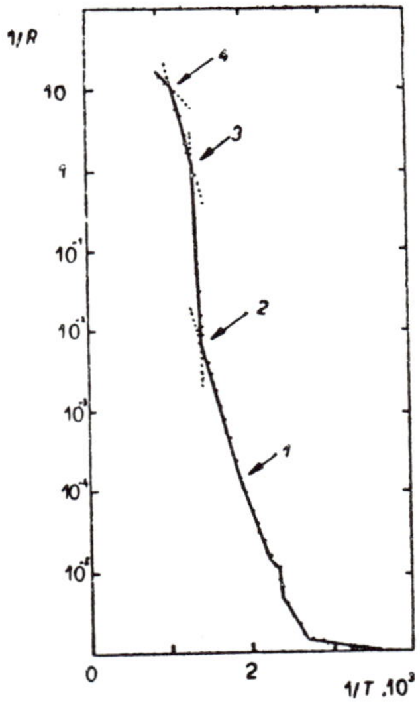

Fig. 5 Temperature dependence of the electrical conductivity.

at 415°C and the abrupt increase in conductivity by two orders of magnitude can be attributed to the separation of metallic silver. Of course, the porosity decrease due to sample shrinkage also contributes to the increase in conductivity. At 460°C, all the silver has been liberated, and the further increase in conductivity with increasing temperature can again be traced back to the semiconductor character. However, the conductivity has changed to the N type. The change in resistance with temperature is reversible, as long as the temperature does not exceed 630°C. This temperature is determined by alternate heating of a tablet to ever increasing temperatures and measurement of electrical conductivity after cooling to room temperature.

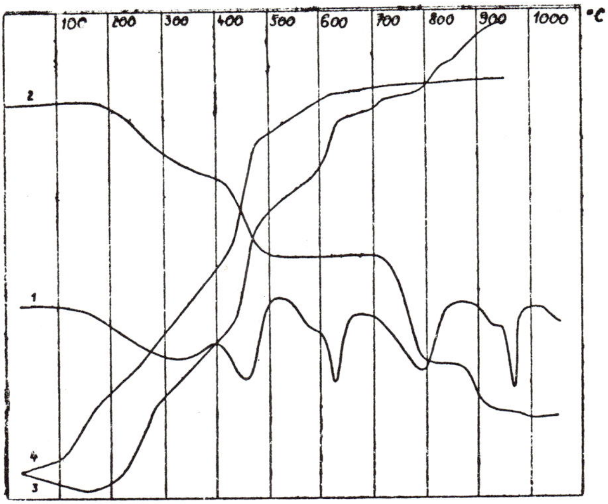

Fig. 6 Changes on heating a product of isothermal decomposition of $AgMnO_4$. 1: DTA curve; 2: TGA curve; 3: contraction thermal analysis curve; 4: curve of electrical conductivity.

The individual processes occurring in the decomposition of $AgMnO_4$ are also accompanied by volume changes, as shown by the contraction thermal analysis curve in Fig. 6. The minimum at 630°C on the DTA curve also involves contraction, although it is not accompanied by a weight loss.

The specific surface area of a product heated to 500°C for 30 minutes was measured at 1 sq. m/g by absorption of butane.

DISCUSSION

The experimental results do not confirm the existence of a primary structural basis of composition 8 $Ag_2O \cdot 15\ MnO_2 \cdot MnO_3$ (modified only by the presence of inactive oxygen), which Hein [3] assumed to be present in the decomposition product of $AgMnO_4$. The empirical formula of this material would be $AgMnO_{2.57}$. However, the TGA curve shows no break that would correspond to this composition, although it has been proved [1] that inactive oxygen escapes most readily from the decomposition product. The existence of ordered structural domains could not be proved by X-ray methods, and the chemical analysis performed by Hein [3] is not sufficient proof for the presence of a definite compound. The decomposition scheme proposed by Grant and Katz [5], which assumes decomposition of $AgMnO_4$ in several consecutive stages, cannot be considered correct either. The existence of the compound $AgMnO_3$ (i. e., silver(II) permanganite), which is supposed to be formed in the first phase of the decomposition process, is not adequately proved, and, on the contrary, is extremely doubtful. There is also no reason to assume that the AgO formed in the second phase must immediately be decomposed further to Ag_2O, since we have shown that AgO can exist up to 400°C in a mixture with MnO_2. The authors evidently made this assumption only to do justice to the finding by Sieverts and Theberath [2] that the decomposition product of $AgMnO_4$ does not form a brown solution of a Ag(II) salt with concentrated HNO_3. However, this fact in no way precludes the presence of AgO, since it follows from our experiments that an artificially prepared mixture of AgO and MnO_2 does not form a solution of this type with concentrated HNO_3 either. The assumption of the presence of AgO in the decomposition product of $AgMnO_4$ was made by Gorgeu [1], in order to explain the source of the inactive oxygen. This assumption is no longer justified, since we have found that the decomposition products of other permanganates also contain inactive oxygen, that the inactive oxygen escapes even at relatively low temperatures, and, finally, that the AgO in the synthetic mixture with MnO_2 is stable up to considerably higher temperatures.

Our experimental results are best explained by assuming that the primary decomposition of $AgMnO_4$, like that of other permanganates [18], proceeds at the MnO_4^- anion, which is transformed chiefly to the MnO_3^- anion. However, this reaction is not completely clear-cut. Different influences, such as the decomposition temperature, type of execution, crystal size, heating time, presence of external catalysts (water vapor, etc.), affect not only the reaction rate [2, 4] but also the composition of the products.

Instead of the permanganite, a certain amount of a mixture of Ag_2O and MnO_2 can also be formed, as indicated by the DTA curves of some glow-decomposition products and of the product of the autoclave decomposition, in which the minimum at 480°C is split into two minima, at 420 and 480°C. We attribute the former to Ag_2O. In view of the fact that a small amount of manganese goes into solution when the decomposition products are extracted with HNO_3, decomposition as far as Mn(III) compounds must be assumed, at least in traces.

All of the products contain more oxygen than corresponds to the overall composition of silver permanganite. Part of this oxygen is dissolved in the solid material (inactive oxygen), part is bonded as manganate, and a further (the smallest) part, which causes the P-type conductivity, is located in interstitial lattice positions.

The differences in composition of the decomposition products obtained in different ways reveal themselves in the course of oxygen evolution during further heating. It is to be assumed that first the adsorbed oxygen, then the oxygen bonded to the manganate anion, and finally the oxygen from the interstitial positions will be evolved. For a continuously increasing temperature, the evolution of this oxygen is extended to higher temperatures, and overlaps with the onset of decomposition of the permanganite, so that no definite zone of stability for the composition $AgMnO_{2.5}$, which corresponds to the permanganite, can be read off on the TGA curve.

The decomposition of the permanganite is complete at 480-500°C, where the decomposition product reaches the composition $AgMnO_2$. This is accompanied by a change in the semiconductor character from the P to the N type, and by an increase of two orders of magnitude in the electrical conductivity. We therefore consider the amorphous material formed to be an atomic dispersion of metallic silver in MnO_2, although it cannot be decided chemically whether it is this or silver manganite-manganese(III) oxide.

The difference between compound and molecular dispersion lies in the assignment of the valence electron to the sphere of silver or of MnO_2. In view of the mobility of these electrons in semiconductors (particularly at high temperatures), the identical behavior of both systems is readily understandable.

The interconversion of the two systems can be expressed by the formula

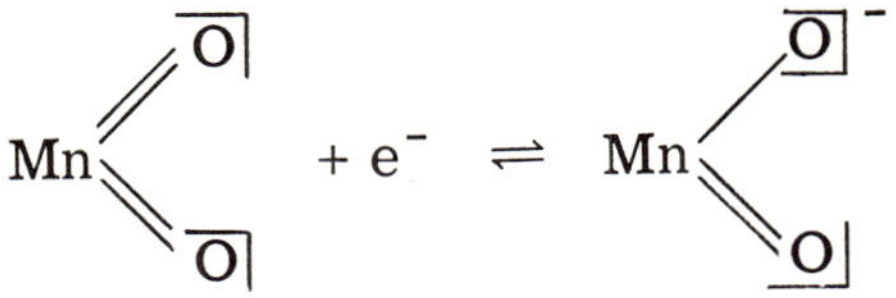

At 630°C, agglomeration of the silver sets in, and the semiconductor character gives way to an essentially metallic conductivity. The decomposition of MnO_2 is displaced upwards up to 700°C; above this temperature, the product consists of silver agglomerates and Mn_2O_3, which is converted to Mn_3O_4 at 940°C.

The catalytic properties of a material of composition $AgMnO_2$ can be explained on the basis of the work of Schwab [19], who showed that the catalytic activity of metals and alloys depends upon their electronic structure. Dowden [20] extended the theory to semiconductors and insulators. In contact catalysis, chemisorption of one or both types of reacting molecules occurs, which is accompanied by a displacement or exchange of electrons between the catalyst surface and the adsorbed molecules, and the mobility of the electron bond in the solid material is therefore very significant.

In the decomposition product of $AgMnO_4$ with composition $AgMnO_2$, the silver atoms represent electron donors, which cause the N-type conductivity, accumulate in the imperfections of the lattice surface, and thus form active centers at which the chemisorption of the reacting materials takes place. In the adsorbed state, a reaction takes place, and the original electronic state of the surface is restored by desorption of the reaction products.

In view of the small specific surface area accessible to the large organic molecules, as shown by the determination of butane adsorption, the high catalytic activity of the material with composition $AgMnO_2$ can be attributed to a high concentration of imperfections at its surface.

SUMMARY

The decomposition of silver permanganate at temperatures up to 1000°C has been followed by differential thermal analysis, thermogravimetric analysis, contraction thermal analysis, and measurements of electrical conductivity and thermoelectric properties. On the basis of the experimental results obtained, a representation of the decomposition processes and of the oxidative catalytic properties of the decomposition product of silver permanganate is given.

LITERATURE CITED

[1] A. Gorgeu, *Bull. Soc. Chim. France* [3] 7, 261 (1892); *Compt. Rend*. **114**, 912 (1892).
[2] A. Sieverts and H. Theberath, *Z. Physik. Chem.* **100**, 463 (1922).
[3] F. Hein, *Z. Anorg. Allgem. Chem.* **235**, 25 (1937).
[4] E. G. Prout and F. C. Tompkins, *Trans. Faraday Soc.* **42**, 468 (1946).
[5] G. A. Grant and M. Katz, *Can. J. Chem.* **32**, 1068 (1954).
[6] E. Moles and M. Crespi, *Anales Soc. Españ. Fís. Quim.* **23**, 198 (1925).
[7] S. Yu. Elovich, S. Z. Roginskiĭ, and E. I. Shmuk, *Izvest. Akad. Nauk SSSR, Otdel. Khim. Nauk* **1950**, 469.
[8] J. Körbl, *Chem. Listy* **49**, 858 (1955); *Collection Czech. Chem. Commun.* **20**, 948 (1955).

[9] J. Körbl, *Chem. Listy* **49**, 862 (1955); *Collection Czech. Chem. Commun.* **20**, 953 (1955).

[10] J. Körbl, *Chem Listy* **49**, 1532 (1955); *Collection Czech, Chem. Commun.* **20**, 1026 (1955).

[11] J. Körbl and K. Blabolil, *Chem. Listy* **49**, 1664 (1955); *Collection Czech. Chem. Commun.* **21**, 318 (1956).

[12] J. Körbl and R. Přibil, *Chem. Listy* **50**, 232 (1956); *Collection Czech. Chem. Commun.* **21**, 322 (1956).

[13] J. Körbl and R. Komers, *Chem. Listy* **50**, 1120 (1956); *Collection Czech. Chem. Commun.* **22**, 222 (1957).

[14] R. Bárta and V. Šatava, *Chem. Průmysl* **3**, 113 (1953).

[15] S. Procházka: "Sammlung von Arbeiten aus dem Gebiet der Technologie der Silikate", Prague (1954), p. 76.

[16] H. Frank: "Semiconductors in Theory and Practice", Prague (1955).

[17] P. H. Emmett, in "Catalysis" (ed. P. H. Emmett), New York (1954), Vol. I, p. 31.

[18] J. Körbl and V. Šatava, Unpublished work.

[19] G. M. Schwab, *Trans. Faraday Soc.* **42**, 689 (1946).

[20] D. A. Dowden, *Research* (London) **1**, 239 (1948).

39

The Micro and Semimicro Carbon-Hydrogen Determination in Fluorine-Containing Compounds with the Aid of the Decomposition Product of Silver Permanganate*

J. Horáček and J. Körbl

In the combustion of F-containing organic compounds, HF is formed under the conditions of elemental analysis, and also SiF_4, by corrosion of the combustion tube. These pass into the absorption tubes, particularly the water absorption tube, and thus cause erroneous results. [1] Belcher and Goulden [2] collect the fluorine in a NaF layer in the fused-quartz combustion tube heated to 270°C., while Throckmorton and Hutton [3] use MgO at 775°C. Mázor [4], who has used a platinum combustion tube for the simultaneous determination of C and H and fluorine, collects the fluorine in a thin layer of Pb_3O_4 in a sintered alumina tube.

In an earlier paper [1], we mentioned the possibility of determining C and H in F-containing compounds, using a long layer of the decomposition product of $AgMnO_4$. [5] With a short filling of the decomposition product, fluorine is not completely absorbed. But even with a long filling, the absorptive capacity is rapidly exhausted in C and H determinations with fluorinated compounds. We have therefore looked for possibilities for a further modification of the original method.

**Collection Czechoslov. Chem. Communs.* 24, 286-90 (1959)

By introducing a layer of pumice impregnated with Pb_3O_4 behind the layer of $AgMnO_4$ decomposition product, the desired goal was achieved at a temperature of about 550° in the catalyst and absorbent filling.

EXPERIMENTAL PARTS AND RESULTS

REAGENTS

Pb_3O_4 on Pumice

Commercial Pb_3O_4 is thoroughly washed with hot, distilled water, and is dried in a drying oven at 105°C. The dry preparation is mixed with ignited pumice (grain size 0.5-1 mm) in the ratio 3:1, in a ground-stoppered flask, and sufficient water is added dropwise for the grains of pumice to be thoroughly coated with Pb_3O_4 during shaking.

Apparatus for Microdetermination

A 10-mm o.d. Supremax combustion tube [1] was extended to 30 cm. The filling, starting from the tip, consists of a 2-cm layer of silver wool, a 3.5-cm layer of Pb_3O_4 on pumice, separated by a short layer of ignited asbestos from the 4-cm layer of $AgMnO_4$ decomposition product, previously ignited for 2 hours at 550°C. The filling is secured in a suitable manner by an asbestos plug. A 10-cm-long electric furnace heats the whole filling to 550°C. Oxygen is obtained from a steel cylinder through a reducing valve, and passes through a pressure regulator and a purifying apparatus of the usual type, filled with Ascarite and Anhydrone. [5, 6] The oxygen flow rate is adjusted to 6-8 ml per min by means of a Mariotte bottle. After filling, the combustion tube is ignited for 20 minutes in an oxygen stream to remove the water contained in the absorbent.

Absorption of Water and CO_2

Pregl absorption tubes [7] are used. N_2O_4, which is formed in the combustion of N-containing compounds, can be absorbed by MnO_2 on silica gel or $Al(OH)_3$ [9] in a tube inserted between the water and CO_2 tubes. With this arrangement it is necessary to heat the entrance of the water absorption tube with a heater. [5]

Apparatus for Semimicrodetermination

This differs from the micro apparatus by a longer combustion tube (35 cm), with extension of the pumice-Pb_3O_4 filling to 5 cm, and the $AgMnO_4$ decomposition product filling to 8 cm. The length of the electric furnace is 15 cm. The pressure regulator is omitted [10], and the oxygen flow rate is adjusted directly to 12-14 ml per min by means of the reducing valve of the steel cylinder. Flaschenträger absorption tubes [11] were used. In the analysis of N-containing compounds, the same tube is used to absorb N_2O_4 after the Pregl water-absorption tube [7] as in the micro method, in order thereby to be able to use the same heater. The apparatus is started as in the preceding case.

PERFORMANCE AND ACCURACY OF THE ANALYSIS

The samples for the microdeterminations weigh from 3 to 4 mg. Solid samples are weighed into platinum boats; liquid samples into capillaries. [6] The sample is burned in 5-7 minutes in the usual way by

TABLE 1

Microdetermination of C and H in F-containing Compounds.

Compound	Formula Mol. Wt.	Calc. Found %C	%H	Diff. Δ%C	Δ%H
2, 2-Dimethyl-3-(*p*-fluorophenyl)-cyclopropanecarboxylic acid	$C_{12}H_{13}FO_2$ 208.22	69.22	6.29		
		69.26	6.52	+0.04	+0.23
		69.11	6.31	-0.11	+0.02
		69.46	6.40	+0.24	+0.11
		69.14	6.31	-0.08	+0.02
(p-Fluorobenzyl)-dimethylcarbinol 3, 5-dinitrobenzoate	$C_{17}H_{15}FN_2O_6$ 362.31	56.35	4.17		
		56.43	4.26	+0.08	+0.09
		56.45	4.28	+0.10	+0.11
		56.43	4.29	+0.08	+0.12
Poly(tetrafluoroethylene)	$(C_2F_4)_n$	24.00	0.00		

		23.88	0.28	-0.12	+0.28
		23.81	0.22	-0.19	+0.22
		23.83	0.27	-0.17	+0.27
3-(*p*-Fluorophenyl)-DL-alanine	$C_9H_{10}FNO_2$ 183.2	<u>59.00</u>	<u>5.50</u>		
		59.13	5.76	+0.13	+0.26
		59.30	5.67	+0.30	+0.17
p-Fluorobenzoic acid	$C_7H_5FO_2$ 140.11	<u>60.00</u>	<u>3.60</u>		
		59.73	3.81	-0.27	+0.21

TABLE 2

Semimicrodetermination of C and H in F-containing Compounds.

Compound	Formula Mol. wt.	Calc. Found % C	% H	Diff. Δ %C	Δ %H
p-Fluorobenzoic acid	$C_7H_5FO_2$ 140.11	60.00	3.60		
		59.97	3.48	-0.03	-0.12
		59.85	3.57	-0.15	-0.03
		59.93	3.66	-0.07	+0.06
		60.03	3.77	+0.03	+0.17
2, 2-Dimethyl-3-(*p*-fluorophenyl)-cyclopropanecarboxylic acid	$C_{12}H_{13}FO_2$ 208.22	69.22	6.29		
		69.25	6.28	+0.03	-0.01
		69.36	6.25	+0.14	-0.04
		69.31	6.34	+0.09	+0.05
		69.32	6.30	+0.10	+0.01
		69.29	6.22	+0.07	-0.07
		69.19	6.22	-0.03	-0.07

Poly(tetrafluoro-ethylene)	$(C_2F_4)_n$	<u>24.00</u>	<u>0.00</u>		
		23.66	0.04	-0.34	+0.04
		23.74	0.18	-0.26	+0.18
		23.82	0.24	-0.18	+0.24
		24.05	0.12	+0.05	+0.12
(*p*-Fluorobenzyl) dimethylcarbinol	$C_{10}H_{13}FO$ 168.2	<u>71.40</u>	<u>7.79</u>		
		71.41	7.74	+0.01	-0.05
(*p*-Fluorobenzyl)di-methylcarbinol acetate	$C_{12}H_{15}FO_2$ 210.24	<u>68.55</u>	<u>7.19</u>		
		68.53	7.12	-0.02	-0.07

means of a gas burner, and the apparatus is rinsed with 100 ml oxygen. The total amount of oxygen passing through the apparatus is 150-160 ml. The absorption tubes are wiped and weighed according to Pregl's directions. [7] A few results are shown in Table 1.

In the semimicro method, the combustion time was 10-15 minutes for 15-20 mg samples, and the apparatus was rinsed with oxygen for 20 minutes. The absorption tubes were weighed 15 minutes after wiping. These results are shown in Table 2.

In attempts to adapt the method for C and H determination to F-containing compounds through the use of the $AgMnO_4$ decomposition product, we have used as absorbents those compounds that have already been used successfully by previous authors for the absorption of fluorine (or HF or SiF_4). [2-4] We selected Pb_3O_4 because this compound absorbs at a temperature that is compatible with that for the catalytic combustion of organic compounds on the decomposition product of $AgMnO_4$ (about 550°C.). This coincidence of temperature permits the use of only one furnace for heating both fillings, as has been described in the experimental part.

The Pb_3O_4 dispersed on pumice quantitatively retains the total fluorine and does not affect the C and H results (see Tables 1 and 2). High H values in the microdetermination have been observed even for fluorine-free compounds. [12] The fluorine absorption proceeds in a limited zone, so that the degree of exhaustion can be followed easily. Depending upon the fluorine content of the organic samples, the absorbent filling can be used for 20-30 semimicrodeterminations.

The use of Pb_3O_4 is not limited to the combination with a support (pumice). However, this combination seemed most advantageous to us, in view of the difficulty in pelletizing Pb_3O_4. Preliminary drying of the absorbent before its introduction into the combustion tube also seems of little advantage, since the Pb_3O_4 particles easily separate from the pumice in the dry state. The chemistry of the action of Pb_3O_4 is dealt with in a paper by Mázor. [4]

The present work again demonstrates the high capacity of the $AgMnO_4$ decomposition product as an oxidative catalyst. This is shown, for example, in the carbon determination of a substance as difficult to burn as polymeric tetrafluoroethylene in whose molecule half of the carbon valences are saturated with fluorine, and where it is impossible to split off fluorine as HF. The compounds listed in Tables 1 and 2 could be burned in the usual manner. Where highly volatile compounds are involved, it is of course possible to apply the special method for the combustion of such compounds together with the correspondingly modified apparatus. [10]

In the analysis of F-containing compounds, the combustion tube corrodes. The tube should be replaced when the corrosion has advanced so far that the combustion process can no longer be followed with certainty. This usually occurs before exhaustion of the catalyst and absorbent filling. In view of the simple construction of the combustion tube, the simple filling procedure, and the rapid breaking-in of the new filling, a replacement involves only a small time loss.

LITERATURE CITED

[1] J. Körbl, *Chem. Listy* **49**, 1532 (1955); *Collection Czech. Chem. Commun.* **20**, 1026 (1955).

[2] R. Belcher and R. Goulden, *Mikrochemie ver. Mikrochim. Acta* **36/37**, 679 (1951).

[3] W. H. Throckmorton and G. H. Hutton, *Anal. Chem.* **24**, 2003 (1952).
[4] L. Mázor, *Mikrochim. Acta* **1957**, 113.
[5] J. Körbl, *Chem. Listy* **49**, 862 (1955); *Collection Czech. Chem. Commun.* **20**, 953 (1955).
[6] M. Jureček: "Organic Analysis", Prague (1950).
[7] F. Pregl: "Die quantitative organische Mikroanalyse", Berlin (1930).
[8] J. Körbl, *Mikrochim. Acta* **1956**, 1705.
[9] J. Horáček, Unpublished results.
[10] J. Körbl and R. Komers, *Chem. Listy* **50**, 1120 (1956); *Collection Czech. Chem. Commun.* **22**, 222 (1957).
[11] B. Flaschenträger, *Z. Angew. Chem.* **39**, 717 (1926).
[12] J. Körbl and K. Blabolil, *Chem. Listy* **49**, 1664 (1955); *Collection Czech. Chem. Commun.* **21**, 318 (1956).

40

The "Isothermal" Decomposition Product of Silver Permanganate as a Combustion Catalyst in the Carbon-Hydrogen Determination*

E. Knižáková and J. Körbl

In previous communications[1-3], we proposed the decomposition product of silver permanganate as a combustion catalyst for both the semimicro-[1,2] and the microdetermination[3] of carbon and hydrogen in organic substances. Recently it has been used also for carbon and hydrogen determination in organosilicon compounds[4] and in fluorine-containing substances[5]. Horáček and Grünberger [6] also use the decomposition product of silver permanganate for the combustion of labeled substances in the determination of ^{14}C. The catalytic properties of the decomposition product of silver permanganate have by now been accurately determined, and the micro method has also proved very satisfactory under statistical evaluation. [7]

However, we occasionally meet the objection that the permeability of the catalyst layer in the combustion tube decreases with time, thus interfering with satisfactory operation of the apparatus. This is caused by the fact that the substance formed by the spontaneous decomposition of silver permanganate[8] is loose, but becomes compressed after a certain

*Collection Czechoslov. Chem. Communs. 24, 2420-21 (1959)

time in the combustion tube. It is clear that difficulties of this type could be easily overcome by pelletizing the decomposition product, but this would remove one of the advantages of using silver permanganate, i. e., its simple, rapid, and easy preparation.

We had found previously in our experiments [9] that the decomposition of silver permanganate can be carried out at relatively low temperatures; i. e., at those too low for deflagration to take place. In the decomposition product thus obtained, the form, appearance, and mechanical properties of the original silver permanganate crystals remain intact. We have now used this "isothermal" decomposition product as a combustion-tube filling for the semimicrodetermination of carbon and hydrogen [2](for absorption of the nitrogen oxides, we have used MnO_2 on silica gel[10]). On the basis of a long series of experiments, the analytical results of the substances analyzed were just as accurate as when using the previously described material. [8] The permeability of the combustion tube remained constant, and no significant compression of the filling was observed. The usual 35-mm catalyst layer, for a combustion tube with i. d. 10 mm, weighs 2-2.5 g, as against approximately 0.8 g when using the deflagrated decomposition product. In this way, the capacity for absorption of interfering halogens and sulfur, and therefore the life of the filling, is also increased.

EXPERIMENTAL PARTS AND RESULTS

PREPARATION OF THE ISOTHERMAL DECOMPOSITION PRODUCT OF SILVER PERMANGANATE

The silver permanganate prepared in a manner

previously described [8], with an average crystal size of 1 × 1 × 3 mm, was heated in a drying oven at 90-95°C for 24 hours in layers no higher than 2-3 mm. The decomposition product formed was then ignited for 1-3 hours at 500°C, and was used directly as a combustion-tube filling in the usual way. Because of its hygroscopic properties, the stock material was stored in tightly closed powder bottles.

ATTEMPTS TO DETERMINE THE CAPACITY

The capacity of a 35-mm catalyst layer, weighing 2.3 g, for chlorine, bromine, and sulfur was determined in the same manner as previously [2], at an oxygen flow rate of 11-12 ml/min. The catalyst layer in question reliably retains 120 mg of sulfur, 240 mg of chlorine, or 1083 mg of iodine.

In contrast to the originally proposed deflagration decomposition product of silver permanganate, the structure of the isothermal product shows no visible change even on absorption of sulfur oxides, so that even in this case the permeability of the combustion tube remains constant.

LITERATURE CITED

[1] J. Körbl, *Chem. Listy* **49**, 862 (1955); *Collection Czech. Chem. Commun.* **20**, 953 (1955).

[2] J. Körbl, *Chem. Listy* **49**, 1532 (1955); *Collection Czech. Chem. Commun.* **20**, 1026 (1955).

[3] J. Körbl and K. Blabolil, *Chem. Listy* **49**, 1664 (1955); *Collection Czech. Chem. Commun.* **21**, 318 (1956).

[4] J. Körbl and R. Komers, *Chem. Listy* **50**, 1120 (1956); *Collection Czech. Chem. Commun.* **22**, 222 (1957).

[5] J. Horáček and J. Körbl, *Chem. Listy* **51**, 2132 (1957); *Collection Czech. Chem. Commun.* **24**, 286 (1959).
[6] J. Horáček and D. Grünberger, *Chem. Listy* **51**, 1944 (1957); *Collection Czech. Chem. Commun.* **23**, 1974 (1958).
[7] M. Večeřa and D. Šnobl, *Chem. Listy* **51**, 1482 (1957); *Collection Czech. Chem. Commun.* **23**, 636 (1958).
[8] J. Körbl, *Chem.* Listy **49**, 858 (1955); *Collection Czech. Chem. Commun.* **20**, 948 (1955).
[9] V. Šatava and J. Körbl, *Chem. Listy* **51**, 27 (1957); *Collection Czech. Chem. Commun.* **22**, 1380 (1957).
[10] J. Körbl, *Mikrochim. Acta* **1956**, 1705.

41

*A Patent for the Use of the Oxidizing Properties of the Thermal Decomposition Product of Silver Permanganate in Catalysis**

J. Körbl

It is known that permanganates, for example, potassium or silver permanganate, are strong oxidizing agents. It appears that in some cases oxidation is due both to the readily available oxygen and to the catalytic effect of the reduction product of the MnO_4^- ion or of the reaction product of the non-dissociated salt in the solid form. It is further known that silver permanganate decomposes with liberation of oxygen when heated to 100°C; it leaves a voluminous residual mass of a black substance insoluble in water. Previously this was considered by some authors to be a mixture of manganese oxide and silver oxide; recent research, however, tends to indicate that the residue is a substance whose composition was expressed by, for example, the formula $8\,Ag_2O \cdot 15\,MnO_2 \cdot MnO_3$. The amount of bound oxygen, however, varies with the mode of preparation; the question of the structure of the compound has thus far not been solved with any degree of finality, and up to now the compound has been of only theoretical interest.

**Czech Patent* 85,465 (1955)

It has now been found that this substance, a decomposition product of $AgMnO_4$ that has been heated to its decomposition temperature, is an outstanding catalytic agent for oxidative reactions and that it retains its activity for a long time when used under sufficiently mild conditions in an oxidative atmosphere. Under such conditions, there is no measurable cleavage of oxygen and the substance acts solely as a catalyst. It is evident that heating in a reducing atmosphere leads to reduction and to a loss of the specific properties of the catalyst.

According to the invention, the catalyst is most applicable in cases where complete combustion of organic compounds is to be brought about in excess oxygen or in air. However, it is also possible to bring about only partial oxidation under carefully regulated mild conditions.

It is possible to activate the catalyst by an admixture of additives such as, for example, compounds of heavy metals, e.g. ZnO, CuO, and others, in order to decrease the level of the threshold temperature of the oxidative reaction. In many instances, however, the substance obtained by thermal decomposition of the pure silver permanganate can be used alone, although it might be necessary to resort to slight heating in some cases in order to initiate the reaction. No further heating is necessary in strongly exothermic reactions; the working temperature of the catalyst is maintained by the heat of the catalyzed reaction.

The catalyst is particularly useful for carrying out oxidations in the gas phase, and it is used in such a manner that the gas mixture of the substance undergoing oxidation and of the oxygen, air or other oxidant is passed through a tube packed with the substance produced by thermal decomposition of $AgMnO_4$. The substance is very fluffy, and it is not necessary to granulate it, since it is readily permeable to gases. Water vapor does not affect the catalyst adversely. The catalyst reacts with halogen, yielding the corresponding silver halide, and with sulfur to form silver sulfite. Nitric oxide is also readily fixed; it reacts with the catalyst even at room temperature to yield silver nitrate. Silver nitrate, as is known, decomposes when heated above 250°C. Heating of the silver nitrate so formed results, contrary to expectation, in reversion of the catalyst to its active form. This means that it is possible to use the catalyst for sorption and desorption of nitric oxide. A practical application, for example, is in the elemental analysis of nitrogenous organic compounds.

According to this invention, the catalyst is suitable for various scientific and technical applications. It was, for example, found very useful in the determination of carbon and hydrogen in organic substances. Combustion can proceed at a much faster rate than usual, and at temperatures which are several hundred degrees below those usually employed. The equipment used therefore shows a much longer service life, and one tube filling suffices for a considerable number of determinations. Interfering halogen and sulfur are held back, since they form the corresponding silver halide or sulfite with the tube filling. The following will, without restriction of the invention to such applications, show a few examples of uses of the new catalyst.

EXAMPLES OF APPLICATION

1. Vapors of organic substances mixed with either oxygen or air are oxidized catalytically by contacting them with the decomposition product of silver permanganate. The decomposition product is heated to between

400 and 450°C to reach its active temperature. Some substances, such as benzoic acid, are oxidized even at much lower temperatures (250°C).

2. When a mixture of carbon monoxide and air or oxygen is passed through a catalyst bed that has been slightly heated, the monoxide is oxidized to carbon dioxide and the contact layer of the catalyst is kept hot by the evolved heat of reaction.

3. A mixture of ammonia and air or oxygen is oxidized on the slightly preheated catalyst to nitrogen oxides. The layer begins to glow, and the glow persists during the entire experiment. The oxides formed can be collected in water in the usual manner.

4. The experiment described in Example 3 was repeated in such a way that air was bubbled through a concentrated aqueous solution of ammonia, and the gas mixture was led (without any preliminary drying) over a bed of the catalyst prepared by thermal decomposition of $AgMnO_4$. The catalyst was preheated. Clear nitric acid, in about 45% concentration, condensed from the effluent end of the catalyst. The acid was completely free of any traces of ammonium salts. The concentration of the acid obtained was found to be a function of the water vapor:ammonia ratio. Because of the elimination of the need for additional oxidation and the easy solution of the nitrogenous gases in water, it is possible to simplify production of nitric acid considerably and to reduce its cost.

The yields are in all cases quantitative, i. e., 100%.

CLAIM

A method of carrying out catalytic oxidation in the presence of excess air or oxygen, characterized in that the material to be oxidized is contacted with the essentially known, low-bulk-density substance formed by heating solid silver permanganate to its decomposition temperature.

42

*The Oxidation of Diphenylamine in the Presence of Silver Permanganate and Its Thermal Decomposition Products**

V. A. Klimova and G. F. Anisimova

In recent years there has appeared in print a series of papers by Körbl [1-5] dealing with the use in elemental microanalysis of both $AgMnO_4$ and the product of its complete thermal decomposition. The author has studied the oxidizing properties of both $AgMnO_4$ and the product of its complete thermal decomposition, i.e., of the product no longer containing $AgMnO_4$. In the combustion of organic compounds with a tenfold excess of $AgMnO_4$, Körbl obtained a quantitative yield of CO_2 and water. In those cases where N-containing compounds were burned, a small amount of nitrogen-containing gases was formed [1]; their yield is not indicated by the author. Körbl has studied in greater detail the oxidative catalytic properties of the product of complete thermal decomposition of $AgMnO_4$. In particular, Körbl reported the quantitative oxidation of ammonia when a mixture of it and oxygen is passed over the heated product of the complete thermal decomposition of $AgMnO_4$ [1].

It was of interest to investigate the extent to which

**Izvest. Akad. Nauk SSSR, Otdel. Khim. Nauk* 1959, 1825-29.

the proposed materials would oxidize the amino group. For this purpose, we have carried out a series of experiments on the oxidation of diphenylamine mixed with $AgMnO_4$ and the product of its complete thermal decomposition. In addition, we have prepared and tested materials obtained by the partial decomposition of $AgMnO_4$. The preparation of $AgMnO_4$ was carried out according to the directions of Körbl, and the decomposition was carried out in a special apparatus which permits following rate and quantity of oxygen evolution.

The thermal decomposition process of $AgMnO_4$ has been widely studied. Decomposition, with evolution of oxygen, starts as low as 100° [6]. The process is autocatalytic in character [7]. The decomposition rate increases at first, reaches a maximum, and then gradually decreases. Upon heating $AgMnO_4$ (100-120°) in our apparatus, we were able to observe the nature of its decomposition as stated above. To obtain the product of complete thermal decomposition, heating was continued until the evolution of oxygen ceased. Partially decomposed $AgMnO_4$ was prepared in two ways: by slow, uniform, partial decomposition and by rapid decomposition. The end of the decomposition in both cases was checked by the weight loss of the $AgMnO_4$ and by the amount of oxygen evolved. We tried to achieve the slow decomposition by using a periodic alternation of heating and cooling of the sample being decomposed. The rapid decomposition was carried out by constant heating, i.e., in the same way as for the complete decomposition of $AgMnO_4$, but, after the decomposition rate reached a maximum, the process was interrupted by rapid cooling. It should be noted that it is not always possible to carry out the decomposition in this way. Because of the autocatalytic character of the process, the decomposition of $AgMnO_4$

frequently proceeds so violently that it is not possible to interrupt it. We have nevertheless succeeded in obtaining materials by both methods that were very similar in composition but different in oxidizing action. The materials obtained by the second method were the more active.

The combustion of a mixture of diphenylamine and the materials obtained showed that the oxidation of the amino group to nitrogen oxides proceeds somewhat differently, depending on the nature of the material added. Thus, the yield of nitrogen oxides is as follows: with $AgMnO_4$, an average of 36% of the theory; with the product of complete thermal decomposition, 25%; with the product of partial thermal decomposition prepared by slow decomposition, 75%; and with the product prepared by rapid decomposition, 85%. In the latter case, the amount of nitrogen oxides rose as high as 96% of theory in individual experiments.

EXPERIMENTAL PART

To study the thermal decomposition of $AgMnO_4$, a glass capsule containing 0.2 g of $AgMnO_4$ was placed in a glass tube with a drawn-out end (Fig. 1). The wide end of the tube was closed with a stopper. The temperature within the tube was monitored with a thermocouple. The drawn-out end of the tube was connected to a Mariotte bottle, and the amount of oxygen liberated was measured by the volume of water flowing out of this. The weight loss was determined by weighing the capsule with the sample. An electric furnace (100-120°) was used for heating. When it was necessary to cool the sample, in order to stop or slow down the decomposition process, the furnace was drawn aside, and a cold compress of wet cotton wool

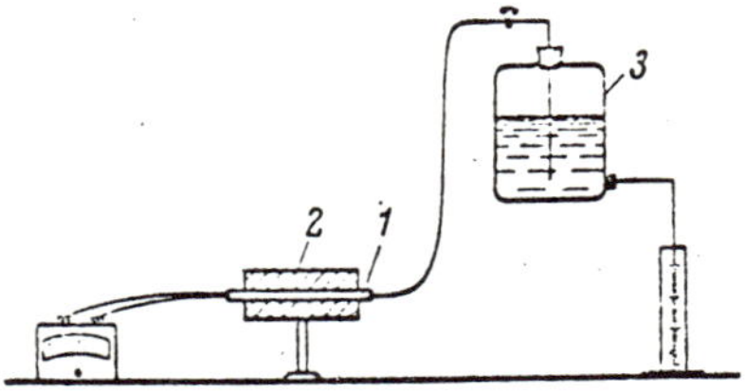

Fig. 1. Apparatus for the thermal decomposition of $AgMnO_4$. 1: glass tube containing a capsule with $AgMnO_4$; 2: electric furnace; 3: Mariotte bottle.

was placed on the corresponding part of the tube. The data obtained are given in Table 1.

It should be noted that the amount of oxygen evolved exceeds the theoretical value by 5-7 ml. The oxygen excess was obtained mainly as a result of expansion of the gas due to heating of the tube and partly as a result of oxygen adsorbed on the surface of the oxidant. The amount of oxygen liberated can therefore serve only as a guide in observing the course of the decomposition. More accurate values are given by the weight loss of the sample studied; however, the adsorbed oxygen can introduce a substantial error in this case as well. Furthermore, in the rapid decomposition, a small amount of the sample may be lost by ejection of sample from the capsule, and we therefore did not always determine the weight loss. Our evaluations of the materials obtained were based mainly upon the analytical data. As seen from Table 1, three types of products of the thermal decomposition of $AgMnO_4$ were obtained; a series of combustions of diphenylamine was carried out with each.

The combustion of diphenylamine was carried out

TABLE 1

Analytical data on the decomposition products of $AgMnO_4$ obtained by various methods

Expt. No.	$AgMnO_4$ (g)	Oxygen evolved (ml)	Weight loss (g)	Analysis* of the decomposition products		Method of decomposition
				$AgMnO_4$	Ag_2O	
1	0.2	19.5	0.018	None	17.7	Heating (110°) until the evolution of O_2 stops.
2	0.2	10.0	0.006	66.6	9.0	Slow partial decomposition (sample is alternately heated and cooled)
	0.2	10.5	0.006	66.2	10.6	
3	0.2	9.0		66.0	6.64	Rapid partial decomposition (110°) (sample is cooled abruptly after going through the maximum rate)
	0.2	9.5		64.1	7.62	
	0.2			53.0		
	0.2			26.2		

*The analysis was carried out according to the directions of Grant and Katz [8].

TABLE 2

Combustion in the presence of $AgMnO_4$

mg Sample	mg Oxidant	mg NO_2	N_2 (% of calc. amount)	Method of combustion
2.740	28	0.300	40.2	In a capsule
3.683	73	0.260	26.0	"
2.880	87	0.340	44.0	"
3.250	130	0.333	37.7	"
3.513	175	0.250	26.3	"
3.800	228	0.348	33.8	"
2.150	230	0.230	39.3	"
2.570	26	0.240	34.3	In a boat
2.812	56	0.350	45.8	"
4.090	122	0.390	35.1	"
2.180	80	0.235	39.3	"
2.820	140	0.170	22.2	"
3.825	230	0.436	45.8	"
3.219	225	0.310	35.5	"

in the apparatus for the rapid determination of carbon and hydrogen [9]. In order to prevent a backward thrust of the combustion gases and their partial absorption by the system serving for the prepurification of oxygen, the empty combustion tube was slightly modified; it had a diameter of 20 mm, length 600 mm, and a spherical widening 60 mm in diameter at its middle. The nitrogen oxides formed were absorbed by a solution of potassium bichromate in 98% sulfuric acid supported on silica gel [10], and were weighed [11]. The weighed sample was taken either in a boat or in a capsule for decomposition. The sample was covered with the material under investigation and was mixed with it by gentle shaking. A stream of oxygen at a rate of 35-40 ml/min was passed through the apparatus during the combustion. The data obtained are shown in Tables 2-5.

TABLE 3

Combustion in the presence of the product of the complete thermal decomposition of $AgMnO_4$.

mg Sample	mg Oxidant	mg NO_2	N_2 (% of calc. amount)	Method of combustion
3.010	30	0.160	19.5	In a capsule
2.800	56	0.232	20.4	"
2.668	80	0.170	23.4	"
3.290	140	0.165	18.4	"
2.870	144	0.285	36.5	"
3.275	198	0.260	29.2	"
2.400	168	0.190	29.1	"

TABLE 4

Combustion in the presence of the product of the partial thermal decomposition of $AgMnO_4$, prepared by slow decomposition.

mg Sample	mg Oxidant	mg NO_2	N_2 (% of calc. amount)	% $AgMnO_4$ in the decomposition product
3.550	110	0.680	74.6	66.6
3.180	120	0.670	77.5	Not analyzed
2.967	83	0.570	70.6	"
2.760	85	0.530	68.9	66.2
2.492	105	0.470	69.3	66.2
2.720	85	0.590	80.0	Not analyzed
4.090	180	0.800	80.2	"
3.090	92	0.530	63.0	66.0

TABLE 5

Combustion in the presence of the product of the partial thermal decomposition of $AgMnO_4$, prepared by rapid decomposition.

mg Sample	mg Oxidant	mg NO_2	N_2 (% of calc. amount)	% $AgMnO_4$ in the decomposition product
3.452	138	0.820	87.3	64.1
3.397	45	0.600	65.0	64.1
2.720	90	0.655	88.5	66.0
3.525	70	0.920	96.0	66.0
3.052	115	0.775	93.3	53.0
2.930	75	0.640	80.3	53.0
3.160	120	0.530	61.5	26.2
2.062	155	0.510	90.9	Not analyzed
1.860	155	0.480	94.9	"
2.800	140	0.710	93.2	"

From the data presented, it is evident that the largest amount of nitrogen oxides is obtained in the combustion when using the partial thermal decomposition product of $AgMnO_4$ which is obtained by rapid decomposition.

SUMMARY

1. Products of the partial thermal decomposition of $AgMnO_4$ have been prepared. The material obtained shows a stronger oxidizing action than $AgMnO_4$ itself or the product of its complete decomposition. The combustion of diphenylamine in its presence proceeds more smoothly than with $AgMnO_4$.

2. It has been shown that the amount of nitrogen oxides obtained in the combustion of diphenylamine with $AgMnO_4$ and its thermal decomposition products varies from 25 to 96%, depending upon the nature of the material used, a fact which is apparently explained by differences in catalytic action of the materials tested.

LITERATURE CITED

[1] J. Körbl, *Collection Czechoslov. Chem. Communs.* **20**, 948 (1955).

[2] J. Körbl, *ibid.* **20**, 953 (1955).

[3] J. Körbl, *ibid.* **20**, 1026 (1955).

[4] J. Körbl and K. Blabolil, *ibid.* **21**, 318 (1956).

[5] J. Körbl and R. Přibil, *ibid.* **21**, 322 (1956).

[6] S. Yu. Elovich, S. Z. Roginskiĭ, and E. I. Shmuk, *Izvest. Akad. Nauk SSSR, Otdel. Khim. Nauk* **1950**, 469.

[7] A. Sieverts and H. Theberath, *Z. Physik. Chem.* **100**, 463 (1922).

[8] G. A. Grant and M. Katz, *Can. J. Chem.* **32**, 1068 (1954).

[9] M. O. Korshun and V. A. Klimova, *Zhur. Anal. Khim.* **2**, 274 (1947).

[10] V. A. Klimova and M. O. Korshun, *ibid.* **6**, 230 (1951).

[11] V. A. Klimova and K. S. Zabrodina, *Izvest. Akad. Nauk SSSR, Otdel. Khim. Nauk* **1959**, 582.

CUMULATIVE JOURNAL INDEX
VOLUMES I and II

Zhur. Anal. Khim. (cont'd)

CUMULATIVE AUTHOR INDEX VOLUMES I and II

SUBJECT INDEX